Processes & Communications

by the same author

Manufacturing Technology

other books of interest

Introduction to Workshop Technology
M. H. A. Kempster, CENG, MIMECHE, AFRAES, MIPRODE

Certificate Mechanical Engineering Science
Applied Mechanics
J. D. Walker, BSC(ENG), CENG, FIMECHE

Certificate Mathematics Vol. 1
N. Abbott, BSC(TECH), AFIMA, DIPMATH

Materials for the Engineering Technician
R. A. Higgins, BSC(BIRM), FIM

Engineering Drawing with Problems and Solutions
K. R. Hart, CENG, MIMECHE, AMJIEE

Processes & Communications

M. Haslehurst
DIC CENG MIMECHE MIPRODE

Head of the Department of Engineering
Warrington Technical College

The English Universities Press Ltd

ISBN 0 340 16263 5 (boards edition)
       0 340 16264 3 (paperback edition)

First printed 1973

The English Universities Press Ltd
St Paul's House Warwick Lane London EC4P 4AH

Printed in Great Britain by
Fletcher & Son Ltd, Norwich

Preface

This book covers the subject matter in the syllabus for the topic of Workshop Processes and Communications at first year (O1) level in the Ordinary National Certificate in Engineering. The syllabus is designed to encourage the study of a limited range of production processes in such a way as to link them with drawing, which is the language of manufacture.

No attempt has been made in this book to impart the skills of draughtsmanship or the skills of the workshop to the reader, but rather to show the relationship between drawing and specification, and its effect upon workshop methods. I have tried to pitch the text at a level suitable for the O1 student, bearing in mind that the great majority of students entering a first year ONC course do so directly from school, or from a suitable college preparatory course, such as the General Course in Engineering. SI units have been adopted throughout the text. Diagrams have been simplified so that their technical content is easily assimilated.

From time to time the student's attention is drawn to the important question of engineering economics. The aspiring engineer has to realise that in a competitive world, no matter how effective and desirable the end product, it will not sell profitably if it has cost too much to make. Also, emphasis is laid on the desirability of safe-working practices. It is my hope that this textbook will help engineering students to pass the O1 level of the ONC course, leading them on to the possession of a worthwhile engineering qualification.

I gratefully acknowledge permission given by the following educational bodies to reproduce questions from their O1 examination papers in Workshop Processes and Communications:

The Welsh Joint Education Committee (WJEC),
The East Midland Educational Union (EMEU),
The Union of Lancashire and Cheshire Institutes (ULCI),
The Northern Counties Technical Examinations Council (NCTEC),
The Union of Educational Institutions (UEI).

The conversion to SI units, where questions were originally set in Imperial units, and the choice of preferred metric basic sizes was undertaken solely by the author.

Extracts from BS 308:1972, 'Engineering drawing practice', are reproduced by permission of the British Standards Institution, 2 Park St, London W1A 2BS, from whom copies of the complete standard may be obtained.

Finally, I would like to thank all my friends and colleagues who have assisted me in any way in the preparation of this book. As was the case with my first book, I owe a particularly heavy debt to my wife, Gwyneth, both for her help in typing the manuscript and for her continual encouragement.

M.H.   Knutsford

Contents

Processes & Communications

Chapter 1
Machining processes

1 Drawing machined components

Ideas may be expressed and transmitted to another person through the medium of words, when the spoken word becomes the means of *communication*. The rules of the language used must be understood by both parties in order for accurate communication to take place. In the same way, ideas may be exchanged between engineers using the medium of a diagram (or more formally a machine drawing) as a means of communication. Similarly the rules of the *language of drawing* must be understood in order that an idea may be accurately expressed.

An engineering designer will use an *engineering drawing* as a system of communication to the engineer responsible for manufacture, to transmit exact instructions about a part or parts which must be made. The drawing language to be used is contained in the British Standard concerned with engineering draughtsmanship, viz. BS 308: 1972, 'Engineering drawing practice'. This standard should be used in order that engineering drawing conventions may be *standardised*

with regard to layout, spacing, relationship of views, sections, dimensions, etc. It will be seen later that what is shown on the drawing will greatly influence the method used for the production of the part(s) and hence the total cost of manufacture. Whether the quantity of parts being made is small ('one off' or a few only) or large, an accurate drawing is necessary in order that unambiguous instructions are transmitted and accurate work results.

Consider an example of a drawing of a component which is required to be produced by machining, i.e. to be shaped to final form and size from a solid piece of metal by a machine tool. Such a drawing is shown in fig. 1.1, and at this stage of our discussion it has been deliberately simplified and in some respects is incomplete. An example of a complete working drawing will be considered later.

Figure 1.1 shows an example of a drawing of a component set out in accordance with the conventions contained in BS 308. (May we at this point urge the reader to look at copies of Standards, and to regard it as part of his education to understand them.) The following points should be considered with respect to fig. 1.1.

a) PROJECTION Three views are shown in this case (the names of the views are shown in brackets, although this is not necessary in practice). In general, the number of views should be the minimum necessary to ensure that there will be no misunderstanding, and they are projected across the drawing either horizontally or vertically, the projection lines not being shown. Ample space is provided between the views.

The drawing in fig. 1.1 is constructed in *first-angle projection*, each view therefore showing what would be seen by looking on the far side of an adjacent view. Alternatively, *third-angle projection* may be used, in which each view shows what would be seen by looking on the near side of an adjacent view (see, for instance, fig. 2.15).

b) DIMENSIONS These are all shown in accordance with BS 308. All hole centres have been dimensioned from face X, face Y, or face Z. In effect, these faces then become the datum planes (the word 'datum' literally meaning 'a given fact') and will be the reference faces from which the hole centres will be located and checked. We will return to this point in sections 3.4 and 4.3. It can be seen, then, that the manner in which a part is dimensioned can affect the methods used for its manufacture and measurement.

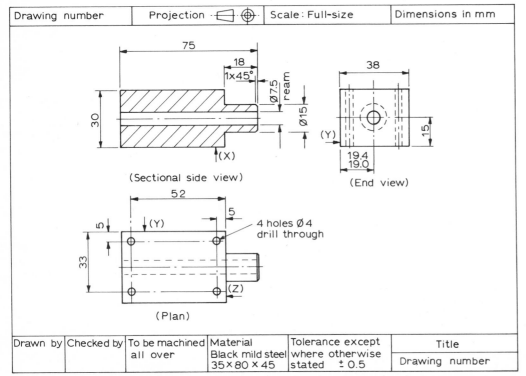

Fig. 1.1 Drawing of a machined component

c) TOLERANCES The *tolerance* is the amount of variation which can be tolerated, to allow for the inherent inaccuracies of the manufacturing process. It is the difference between the *high limit* (which is the largest permissible dimension) and the *low limit* (which is the smallest permissible dimension). Figure 1.1 shows a part with a general tolerance of ±0·5 mm, excepting the 19 mm dimension, which means that all other dimensions are allowed a maximum variation of 1 mm. Again, the tolerance inevitably affects the chosen method of manufacture. A small ('tight') tolerance implies that a more accurate machining process must be used, to be followed by a more accurate inspection method; wider tolerances imply the opposite. Therefore, it can be seen that the *choice of tolerance*, which affects the manufacturing method, will in turn influence the *cost of production*. The imposition of unnecessarily tight tolerances inevitably increases the cost.

There is a close relationship between the specified tolerance and the surface finish which will result from the machining process (not referred to in fig. 1.1), as also there is between tolerance and the type of fit between assembled parts. Surface finish will be discussed in section 3.3, and limits and fits in section 3.2.

d) MATERIAL The choice of *material* must be specified on the drawing, and in practice it is not sufficient just to state 'black mild steel', for instance, as there are several such such steels available. A particular material specification should be referred to, such as BS 970. This is an important standard which describes the whole range of wrought steels in the form of bars, billets, and forgings; it will be referred to in more detail in Chapter 6.

The component shown in fig. 1.1 will be made from a piece of 'black' M.S. sawn from a bar; This is then to be machined all over. Alternatively, it might have been possible to specify 'bright' M.S., having a better finish which might have proved satisfactory for the flat surfaces. Less metal removal would have been necessary (cost saving), but the bright material would have been more expensive (cost increase). Now it can be seen that the *choice of material* may also affect the production method and again the *cost of production*.

1.2 Link between design and production
Figure 1.2 illustrates that the drawing links the functions of design and manufacture. A working drawing is in a sense a

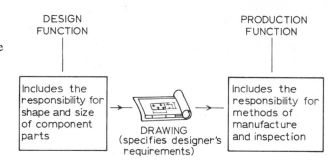

Fig. 1.2 Drawing linking design and production functions

specification outlining the designer's requirements by means of the language of drawing. A specification can be defined as a *detailed description*. We have already seen that the dimensioning, tolerance, surface finish, and material specified can affect the method and ultimate cost of production; it would therefore be an ideal situation in industry if designers were experts in production processes. Unfortunately this is not always so, in which case a link (liaison) man may be required–a product engineer well versed in the technology of manufacture who is also capable of modifying designs in order to make the production easier, faster, and cheaper (all combined in the term 'more efficient'). This product-methods engineer should work with the designer at the drawing stage, later collaborating with all concerned in the planning, tooling up, manufacture, and inspection of the product.

Let us consider one or two of the problems concerned with the manufacture of the designed part shown in fig. 1.1. The quantity required will largely determine the types of machines and tools used, and it will be appreciated that, even in making one off, several acceptable alternative methods may be available. Fortunately, some basic principles hold good whatever the circumstances: in the case of machining, the most satisfactory method will be that which involves the least number of set-ups, hence reducing the chances of error (due to setting-up) to a minimum. We will assume that the sequence of operations shown in fig. 1.3 has been decided upon.

Note that it is in order for the designer to specify written machining instructions alongside a dimension, such as '$\phi7·5$ ream', or '4 holes 4mm drill thro'. Obviously such instructions dictate the production method as surely as do specified tolerances, surface finish, etc.

Now, say that after operation 1 the block of metal is on the lower limit, and it is then set up in the lathe such that the axis of the reamed hole lies in the centre of the 38 mm dimension, apparently 38/2 or 19 mm away from face Y.
In fact, of course, the reamed hole axis will measure 37·5/2 or 18·75 mm from face Y, and this is outside the allowed limits. This illustrates the importance of locating features in the correct position from datum faces as called for on the drawing. In this case the reamed hole axis must be positioned 19·0 to 19·4 mm from face Y – if it were required to be exactly in the centre, then the drawing would have to be dimensioned differently.

The same principle applies to the 4 mm diameter holes, which need to be positioned 5 mm and 32 mm away from face Y and 5 mm and 52 mm away from face Z. Whether one or a large quantity is required, it will be necessary to position (locate) the part in the correct relationship to the drill(s) at the setting stage, prior to machining. Figure 1.4 illustrates the principle.

Opn. no.	Sketch of operation	Description of operation	Machine
1	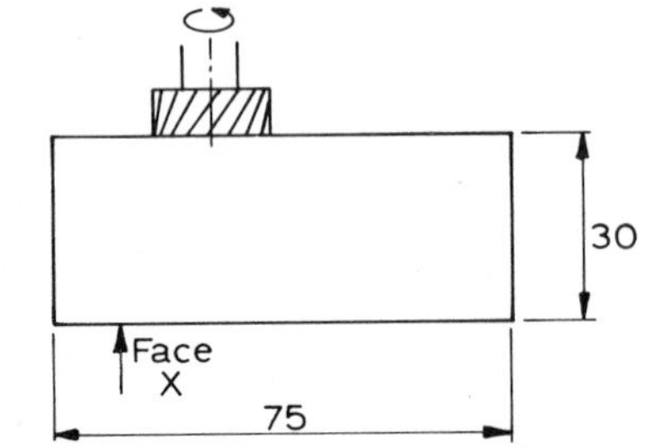 Face mill all over to finished size	Vertical miller
2	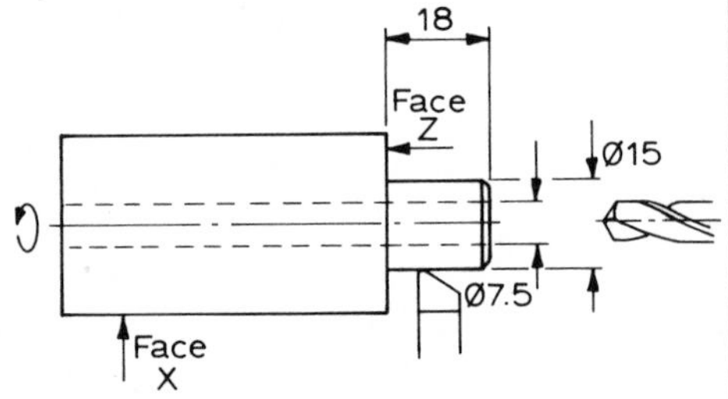 Turn & chamfer boss Centre drill Drill Ream	Lathe
3	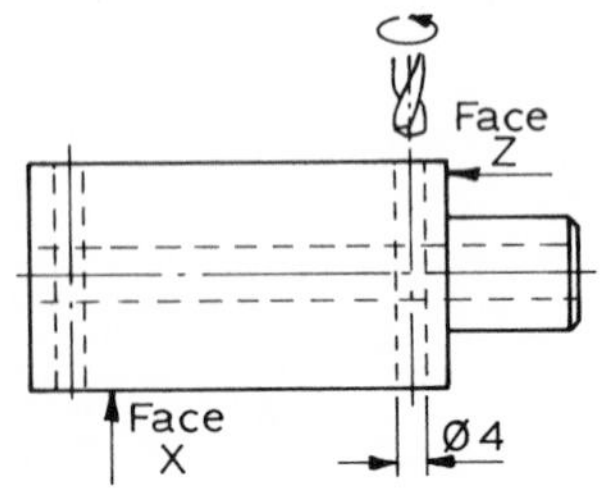 Drill 4 holes	Vertical drilling machine
4	Inspect	

Fig. 1.3 Manufacturing sequence of operations

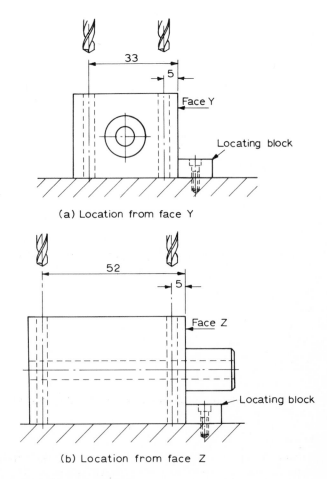

(a) Location from face Y

(b) Location from face Z

Fig. 1.4 Locating drilled holes

4

Hence it is clear that the drawing *specification* largely dictates the appropriate *method* of manufacture. Now consider one other example.

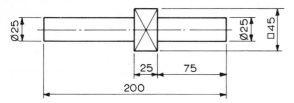

Fig. 1.5(a) Metal component

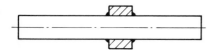

Fig. 1.5(b) Fabricated metal component

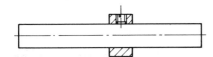

Fig. 1.5(c) Fabricated metal component

Say a shouldered component is required, as drawn in fig. 1.5(a); depending upon the quantities required, surface finish, material specified, and accuracy, the method of production can be determined. It may, for example, be:
a) machined from square or round bar (costly in time and material);
b) cast or forged to shape (followed by machining where necessary);
c) fabricated by welding, brazing, or soldering two pieces together [see fig. 1.5(b)];
d) fabricated by securing two pieces together with a screw [see fig. 1.5(c)];
e) fabricated by joining two pieces together by driving, or by shrinking using heat.

There could be other *methods* of manufacture, all affecting the final *cost*.

The finished detail or working drawing of this part will reflect the designer's wishes, which in turn will be determined by the function of the part. Figure 1.6 shows the finished drawing, from which it is clear which method of manufacture must be adopted.

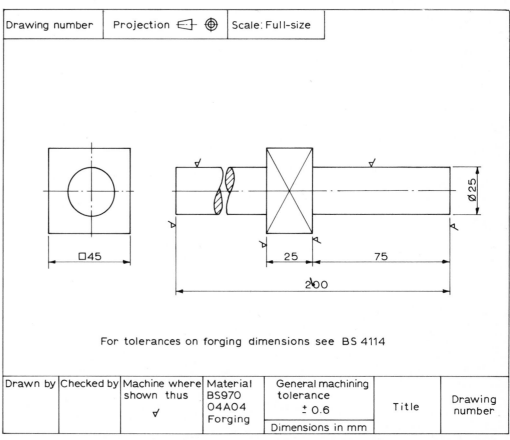

Fig. 1.6 Drawing of a machined component

1.3 Geometric basis of machining processes

The function of a machine tool is to hold a workpiece in the correct location with respect to the cutting tool; then to allow the tool and work to move in the correct *geometric relationship* to each other so that metal is removed from the work, hence giving the required final shape. This final solid object will be a combination of *plane* (flat) *surfaces* and/or *cylindrical surfaces*. (We will ignore more difficult geometric relationships.) It will be seen that such machined surfaces are produced by *straight-line* or *circular-arc* movements of the tool (cutter) or work. Figure 1.7 illustrates the principle for flat surfaces only.

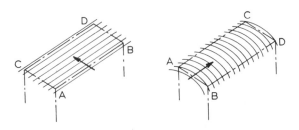

Fig. 1.7 Flat surfaces

Figure 1.7(a) shows the straight line AB advancing to position CD and producing the flat surface shown by chain-dotted lines. Figure 1.7(b) shows the plane circular arc AB advancing to position CD and sweeping out the flat surface again shown by chain-dotted lines. Figure 1.8 illustrates the principle for cylindrical surfaces.

Figure 1.8(a) shows the straight line AB rotating about a straight parallel axis and producing the circular surface shown by chain-dotted lines. Figure 1.8(b) shows the circle AB advancing along a straight line to position CD and producing the circular surface again shown by chain-dotted lines.

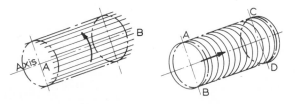

Fig. 1.8 Cylindrical surfaces

Forming and generating

Straight-line or circular-arc profiles can either be *formed* or be *generated* using a single-point tool or revolving cutter. Forming is carried out using a form tool or cutter which bears the required profile upon its cutting edge; the shape produced upon the work will therefore be the reverse image of the tool profile. Figure 1.9(a) shows a circular profile being formed upon a workpiece, the tool or work being reciprocated lengthways. Note that the resultant shape is entirely dependent upon the tool profile.

Fig. 1.9(a) Forming a semi-circle

Generating is carried out by means of the tool or work being constrained to move through the required path, the desired profile therefore being generated upon the work. In this case the tool profile has no effect upon the resultant work shape. Figure 1.9(b) shows a circular profile being generated upon a workpiece, the tool being reciprocated lengthways while the work is rotated.

Much of the machined engineering work is not only a combination of flat and circular surfaces (or more complicated shapes) but also a combination of both formed and generated surfaces. Consider the work shown in fig. 1.9(a); its profile looking at the end elevation is formed, but its shape looking at the side elevation is the result of generating. The straightness of the part along its length is a function of the straightness of the slideways controlling the reciprocating movement of the tool-slide (or work-slide).

To conclude this section we will illustrate some of the common ways of producing both flat and circular surfaces upon a machine tool, leaving it to the reader to determine whether forming or generating is taking place.

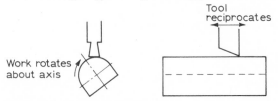

Fig. 1.9(b) Generating a semi-circle

Figure 1.10 shows common methods of machining flat surfaces. Figures 1.10(a) and (b) should be compared with figs 1.7(a) and (b) respectively.

Figure 1.11 shows common methods of machining circular surfaces. Figures 1.11(a) and (b) should be compared with figs 1.8(a) and (b) respectively.

The process of broaching shown at fig. 1.11(b) is a high-speed machining process for producing internal profiles, but will not be considered in detail in this book; however, the basic machining processes of turning, shaping, milling, and drilling will be discussed in the final four sections in this chapter. Only the main principles involved will be considered – descriptions of skills and machine-tool design will be found in other, more specialised, textbooks.

(a) Shaping-tool reciprocates across work-work feeds past tool.

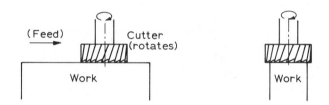

(b) Vertical (face) milling – work feeds past cutter.

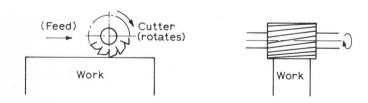

(c) Horizontal (peripheral) milling – work feeds past cutter.

Fig. 1.10 Production of flat surfaces

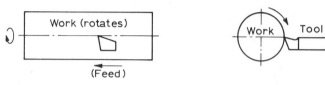

(a) Turning – tool feeds across work

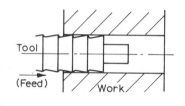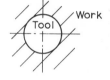

(b) Broaching – tool feeds through work

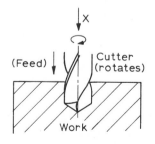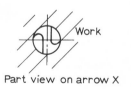

(c) Drilling – cutter feeds through work

Fig. 1.11 Production of circular surfaces

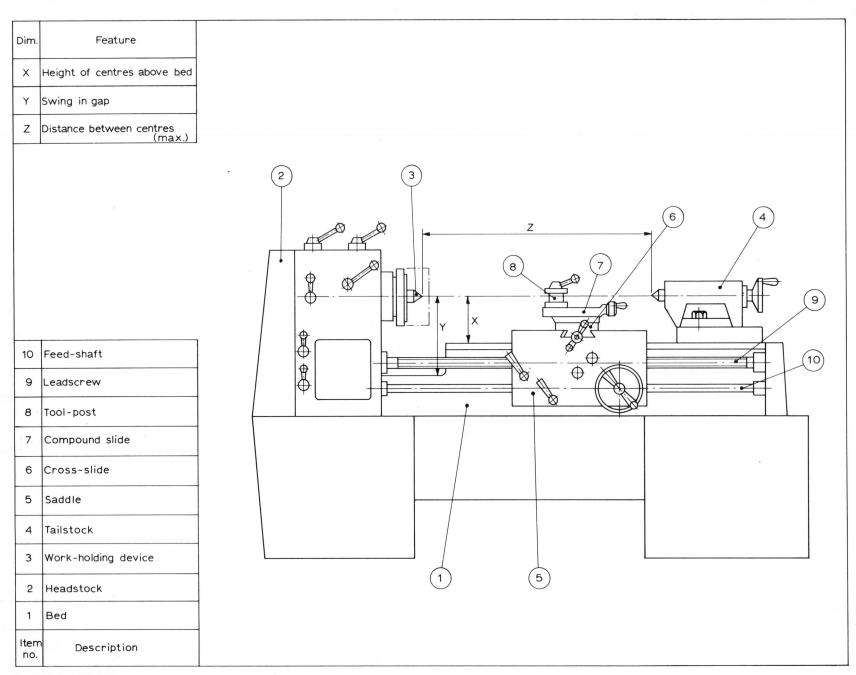

Dim.	Feature
X	Height of centres above bed
Y	Swing in gap
Z	Distance between centres (max.)

Item no.	Description
10	Feed-shaft
9	Leadscrew
8	Tool-post
7	Compound slide
6	Cross-slide
5	Saddle
4	Tailstock
3	Work-holding device
2	Headstock
1	Bed

Fig. 1.12 Centre lathe

1.4 Production of work on a centre lathe

A *centre lathe* is a machine tool used for the manufacture of plain cylindrical shapes or flat surfaces by generation or forming. It is most suitable for small batch production, i.e. ones or tens, but derivatives of it, such as the capstan lathe, are used for quantity production, i.e. batches of 100's or 1000's. The work may be in the form of a bar, billet (sawn piece of bar), casting, or forging, and may be held in a chuck, upon a faceplate, or between centres. Plain turning, taper and form turning, drilling, boring, facing, and threading are the main operations performed on this machine.

Figure 1.12 shows a diagram of a centre lathe. This diagram shows only the major features, with the most important dimensions – X, Y, and Z – added. Centre lathes are usually classified by X and Z dimensions; for example, a 190 mm × 760 mm lathe would indicate that work up to a maximum of 380 mm diameter × 760 mm long could be accommodated between centres.

We will briefly consider each major feature in turn.

1) BED This constitutes the *main structure*, being a casting of box form, well ribbed and cross-braced to prevent deflection and twist. It is made of cast iron of uniform thickness, and may have a gap to accommodate short jobs of larger diameter, as shown. The top of the bed has induction- (electrically) or flame-hardened slideways of inverted vee and/or flat section, accurately ground or scraped, to support the saddle and tailstock.

2) HEADSTOCK On modern machines the headstock is usually of the 'all-geared' type, containing hardened and ground gears giving say 8, 12, or 16 spindle speeds ranging from say 30 to 5000 rev/min. The drive from the electric motor is through a clutch, a brake being included as a safety feature to give fast stopping of the spindle.

3) WORK-HOLDING DEVICE A three-jaw self-centring chuck or a four-jaw independent chuck (with reversible jaws) is used for gripping and driving bar, billet, castings, or forgings. A collet chuck is useful for bar work.

Driving plates are used with work held between centres, in conjunction with a carrier clamped to the work.

Faceplates with clamps are used to hold castings or forgings of unusual shape which cannot conveniently be chucked. The work should be *balanced* by adding weights to the faceplate before machining takes place. This is necessary in the interests of both safety and accuracy.

4) TAILSTOCK This can be bolted in any desired position on the slideways. It carries a centre for supporting work, or it can be used for feeding drills, reamers, etc. into the work. In the latter case it will carry a drill chuck, or drill, reamer, tap, etc. For accurate work to be produced, it is essential that exact alignment is obtained between the tailstock and machine axes, in both the vertical and the horizontal planes.

5) SADDLE The saddle is the main slide upon which are mounted the lesser slides (6) and (7), and it can be either manually or automatically traversed along the slideways. The top portion has a long bearing surface and is termed the *carriage*, the front portion being termed the *apron*. This houses the feed mechanism and operating levers for controlling the feed of the slides.

6) CROSS-SLIDE This may be traversed across slideways machined on the top of the saddle. A graduated dial is fitted to the traverse screw which actuates the slide (see section 3.5).

7) COMPOUND-SLIDE This is mounted on slideways machined on the top of the cross-slide, and carries the tool-post. It takes its name from the fact that it can be set at any angle, hence giving it a variety of positions.

8) TOOL-POST The tool-post is fixed to the compound-slide. It is made in a variety of shapes, but in general is of the single or multi-tool type. The tool(s) is set on the work centre, and then clamped. On some modern machines the tool is fitted in a tool-holder which may be interchanged with other tool-holders on a multi-type tool-post. It may then be replaced when required, without re-setting – this gives great savings in time (see section 2.4.).

9) LEADSCREW This is a master screw usually having a high accuracy Acme thread of 6 mm pitch. The leadscrew may be connected to the saddle by means of a split nut, the nut being closed onto the rotating leadscrew by means of a lever during screw-cutting.

The leadscrew and main-spindle speeds are synchronised by being connected together through gears. The correct gear ratio to give the desired thread pitch upon the work is selected by levers on the outside of the gear-box. The speed at which the leadscrew rotates (in relation to the spindle) obviously determines the speed at which the saddle moves along the bed, this in turn determining the pitch of the machined thread.

10) FEED-SHAFT This is the means by which the saddle is traversed along the bed for plain turning. It is connected to the main spindle through gears in a gear-box, the feed-shaft speed determining the saddle feed-rate–see fig. 1.13.

Figure 1.13 shows diagrammatically the connection between the main-spindle rotation and the saddle linear movement (feed), which can be expressed in units of millimetres per revolution of the main spindle. Therefore, if the feed-shaft is set to run at a particular speed, it can be seen that this speed will increase if the main-spindle speed is increased, the saddle 'feed' remaining the same.

Safety features

Although speed is the essence of production engineering, and time can be equated to cost, *safety* of the operator must be the first consideration. Skill-training on machine tools must always be based upon safe-working practice. Also, modern lathes have built-in safety features, some of the more important being:

a) totally enclosed gear-boxes,
b) totally enclosed motor and vee-belt drives,
c) spindle noses without projections,
d) foolproof electrical switchgear mounted in watertight housings,
e) chuck guards,
f) self-ejecting chuck keys,
g) ductile iron chucks for use at high spindle speeds,
h) low-voltage lighting,
j) 'safe-gate' and interlocking controls,
k) switches and controls conforming to a standard code of marking.

Standard operations

Certain standard settings are used to produce the basic elements of form upon the workpiece, some of which are considered here. In general, the bulk of the work material is removed by a roughing cut(s), the desired shape and size being accurately produced by a final finishing cut.

Figure 1.14 shows plain external-turning operations, and fig. 1.15 shows plain internal-turning operations.

Figures 1.14(a) to (d) and (j) show operations requiring a sliding feed, i.e. saddle traverse. Figures 1.14(e) to (g) show operations requiring a surfacing feed, i.e. cross-slide traverse. Figure 1.14(h) shows an operation requiring a compound-slide feed. Figure 1.14(j) shows a previously bored workpiece pressed onto a tapered mandrel which is rotated between centres–the outside diameter can then be turned concentric to the inside diameter with little difficulty.

If the operations shown at figs 1.15(a) to (d) were carried out in that sequence, then a true hole would result, being geometrically round, straight, axially true, and of high accuracy.

A 'steady' must be used to support long, slender work. Two types are available: the fixed steady, which is clamped to the lathe bed, and the travelling steady, which is fixed to the lathe saddle.

Other accessories available for lathes, which are particularly useful for repetition work, include taper-turning attachments, copying attachments, and six-station bed turrets.

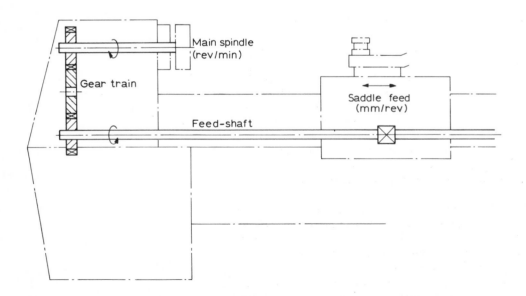

Fig. 1.13 Feed-shaft gearing

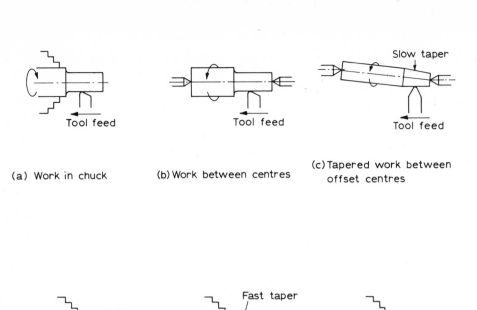

(a) Work in chuck

(b) Work between centres

(c) Tapered work between offset centres

(d) Threaded work in chuck

(e) Tapered work ('plunge' cut)

(f) Parting off work from bar

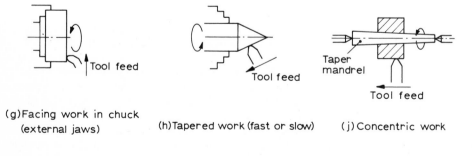

(g) Facing work in chuck (external jaws)

(h) Tapered work (fast or slow)

(j) Concentric work

Fig. 1.14 External turning

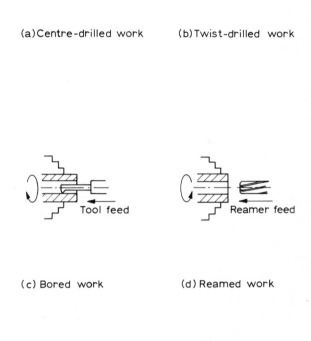

(a) Centre-drilled work

(b) Twist-drilled work

(c) Bored work

(d) Reamed work

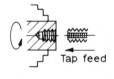

(e) Threaded work

Fig. 1.15 Internal turning

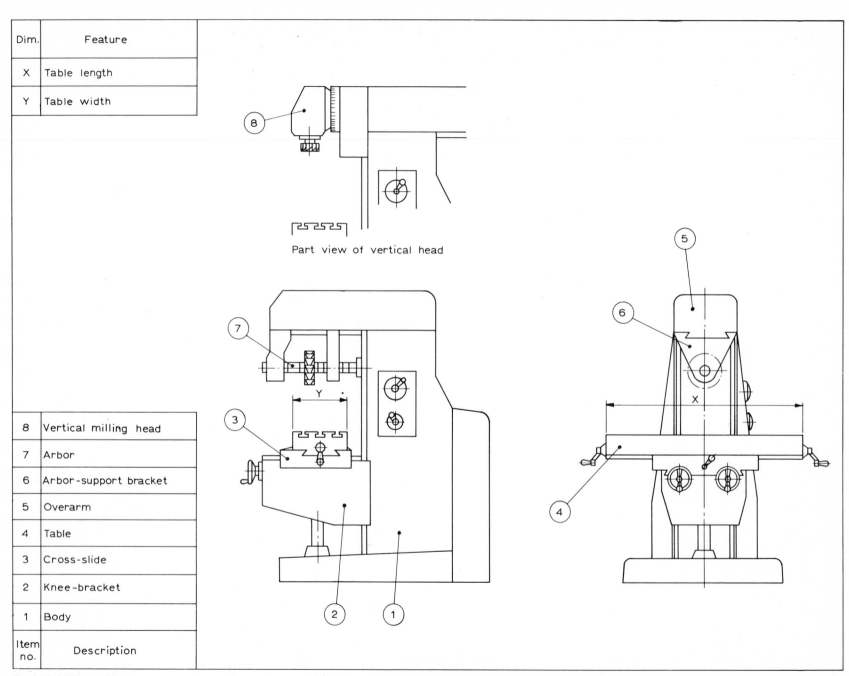

Dim.	Feature
X	Table length
Y	Table width

Part view of vertical head

8	Vertical milling head
7	Arbor
6	Arbor-support bracket
5	Overarm
4	Table
3	Cross-slide
2	Knee-bracket
1	Body
Item no.	Description

Fig. 1.16 Milling machine

1.5 Production of work on a milling machine

A *milling machine* is a machine tool used for the manufacture of flat or profiled surfaces by generation or forming. Like the centre lathe, it is most suitable for small-batch production, but derivatives of it, such as the fixed-head machine, are specialist quantity-production machines. The work may be in the form of bar, billet, casting, or forging, held in a vice, clamped upon the table, or mounted in a dividing (indexing) head. *Horizontal* machines utilise cylindrical cutters having teeth on the periphery, and *vertical* machines utilise cylindrical cutters mainly using teeth on the face to remove workpiece material – horizontal and vertical referring to the cutter axes. *Universal* machines are used for either function, and may be said to be in the same class as centre lathes for versatility.

Figure 1.16 shows a diagram of a horizontal milling machine, with the vertical configuration indicated above the horizontal configuration. Universal machines offer both facilities, with the addition of swivelling arrangements for the table.

The diagram shows only the major features, with the most important dimensions, X and Y, added. For example, a 1000 mm (X dimension) $\times$ 250 mm (Y dimension) milling machine would indicate that work up to a maximum of 1000 mm long $\times$ 250 mm wide could be accommodated on the table.

We will briefly consider each major feature in turn.

1) BODY The body is the *main structure* of the machine, made from a specially strong cast iron, and incorporates the vertical column upon whose prismatic slideways the knee-bracket slides. (Note – 'prismatic' means 'appertaining to a prism', which is a solid whose ends are equal plane figures and whose sides are parallelograms.)

Prismatic slideways are used on modern machines instead of the large, flat ways favoured on earlier models.

2) KNEE-BRACKET This is usually a casting of massive proportions upon whose top face are slideways to accommodate the cross-slide (alternatively called the saddle). It can be traversed vertically up or down, and has a tubular support underneath.

3) CROSS-SLIDE This carries the work-table in slideways, and can itself be traversed, giving the cross-ways movement to the work.

4) TABLE The table is made from a high-class cast-iron alloy, being machined on all surfaces. It can be traversed in the cross-slide ways, giving longitudinal movement to the work. Tenon slots machined in its top face accommodate 'tee-bolts' which are used to hold clamps, vices, or dividing heads upon the table.

5) OVERARM This may be of round or rectangular section, as drawn. On universal machines it may be adjusted to the required position. It carries the arbor-support brackets, which may be adjusted along its ways.

6) ARBOR-SUPPORT BRACKET This is fitted with a phosphor-bronze bearing bush to support the arbor at its free end. An additional bracket can be fitted when required, as shown in the diagram.

7) ARBOR (see section 2.4) The arbor is a plain spindle onto which the cutter or cutters (collectively called a 'gang') are placed, then clamped between collars. The arbor obtains its drive from the taper bore in the main-spindle nose, into which the arbor fits. This spindle nose can also accommodate chucks or cutters directly, instead of the arbor. The spindle is driven through nickel-chrome steel gears, the desired speed being selected from the side of the column. An electric motor transmits the drive to the gears through the medium of a friction clutch, a brake being used to give safe, fast spindle (and hence cutter) stopping.

8) VERTICAL MILLING HEAD This can usually swivel through 360° in the vertical plane, and the spindle-speeds range is the same as for the horizontal arbor. The spindle nose can accommodate chucks or cutters.

TABLE-FEED ARRANGEMENTS It should be noticed, in this case, that the table, unlike the lathe saddle, has its own feed gear-box, which can be used independently of the main spindle; therefore, any main-spindle speed alteration will not affect the speed (feed) at which the table traverses. The table movement, then, can be expressed in units of mm/min, as can the feeds of the cross-slide and knee-bracket.

Safety features

It is the author's opinion that a milling machine, particularly the horizontal version, is potentially one of the most dangerous machines in an engineering workshop. It

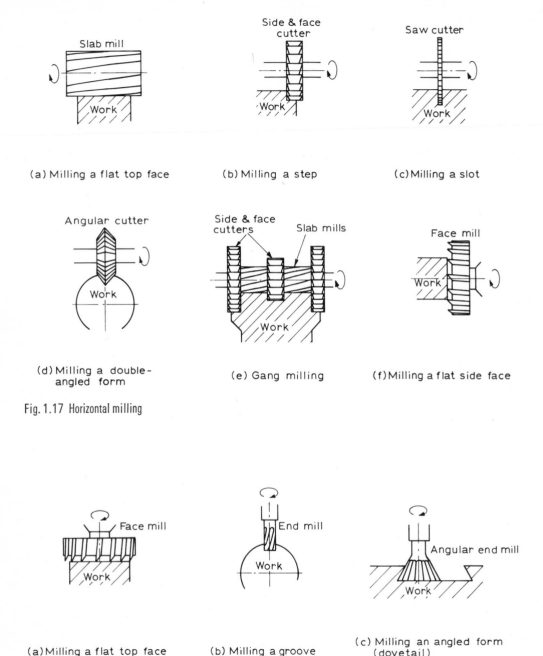

(a) Milling a flat top face (b) Milling a step (c) Milling a slot

(d) Milling a double-angled form (e) Gang milling (f) Milling a flat side face

Fig. 1.17 Horizontal milling

(a) Milling a flat top face (b) Milling a groove (c) Milling an angled form (dovetail)

Fig. 1.18 Vertical milling

must never be used without guards which completely enclose the cutter(s), giving foolproof protection.

Many of the built-in safety features are similar to those used upon centre lathes, additional features being:

a) slipping clutches for all feed motions,
b) hand and automatic feed levers interlocked,
c) limit stops to prevent over-traversing.

Standard operations

Certain standard settings are used to produce the workpiece shape, some of which are considered here.

Figure 1.17 shows plain horizontal-milling operations, and fig. 1.18 shows plain vertical-milling operations.

Figures 1.17(a) to (d) show single cutters mounted on the arbor. Figure 1.17(e) shows a gang of cutters mounted on the arbor, giving higher productivity. Note the opposite helices of the slab mills to equalise the end cutting thrust. A simpler variation of this is called *straddle milling*, in which two side-and-face-cutters are used to machine opposing parallel faces. Figure 1.17(f) shows a cutter mounted directly on the spindle nose, the arbor not being used.

Depending upon the batch quantity and its shape and size, work may be clamped directly to the table, held in a vice, or mounted in a chuck or between the centres of a dividing head. For production work, where large quantities, are required a *fixture* may be used. This is a device which locates and holds the workpiece(s) such that it is positioned in the correct relationships to the cutter(s). Fixtures are used on production lathes and milling machines, for example. The design of fixtures represents an important part of the work of a tool-drawing office.

It will be seen from figs 1.18(a) and (b) that these operations could equally well be horizontally milled, but that operation (c) could not be horizontally milled. The many alternatives available in engineering processes often prove confusing to the young practitioner, and therefore the working principles should be studied at some length.

It is probably true to say that *vertical face milling* is used in preference to horizontal milling on most production lines, particularly on first operations on castings or forgings, where the cutter teeth contact a hard scale. The area of contact is much less on a face mill; in addition, the vertical process is quicker, smoother, uses less power, and produces a superior finish. Face mills lend themselves to the fitting of inserted teeth, usually of tungsten carbide.

1.6 Production of work on a shaping machine

The *shaping machine* is used mainly for the manufacture of flat surfaces, being relatively simple in conception and quick to set up. Because of its reciprocating configuration, giving an idle reverse stroke of the tool, it has little place on a production line – the milling machine being preferred. Derivatives of it are the slotting machine, which may be regarded as a vertical shaping machine, and the planing machine, which accepts larger work.

The types of work and work-holding devices are similar to those encountered in the milling process, with the exception of fixtures, which are rarely used for shaping. The cutting tools are single-point, similar to those used for turning, but of more robust section to withstand greater cutting forces. Carbide-tipped tools are frequently used, because much shaping work consists of the first roughing operation on castings or forgings. Figure 1.19 shows a diagram of a shaping machine.

The diagram shows only the major features, with the most important dimensions – X, Y, and Z – added. Usually, shaping machines are referred to in terms of size by means of the ram stroke, i.e. a 450 mm machine would mean that a maximum stroke-length of 450 mm was available.

All modern machine tools can be obtained with either mechanically or hydraulically actuated mechanisms. Hydraulic machines are generally more expensive, but have the advantages of simplicity, quietness in operation, and infinitely variable speed- and feed-changes. Figure 1.19 shows a hydraulic shaping machine with the reciprocating ram controlled by a hydraulic cylinder. The major features are as follows.

1) BODY This is a massive casting of box section, well ribbed for strength and rigidity. It incorporates the base and the column, the latter having vertical slideways at the front, to accommodate the main slide, and horizontal slideways along the top, to accommodate the ram.

2) MAIN SLIDE This can be adjusted for position in the vertical slideways of the body by means of the elevating screws mounted underneath, on the base. It has long horizontal slideways in which the table moves.

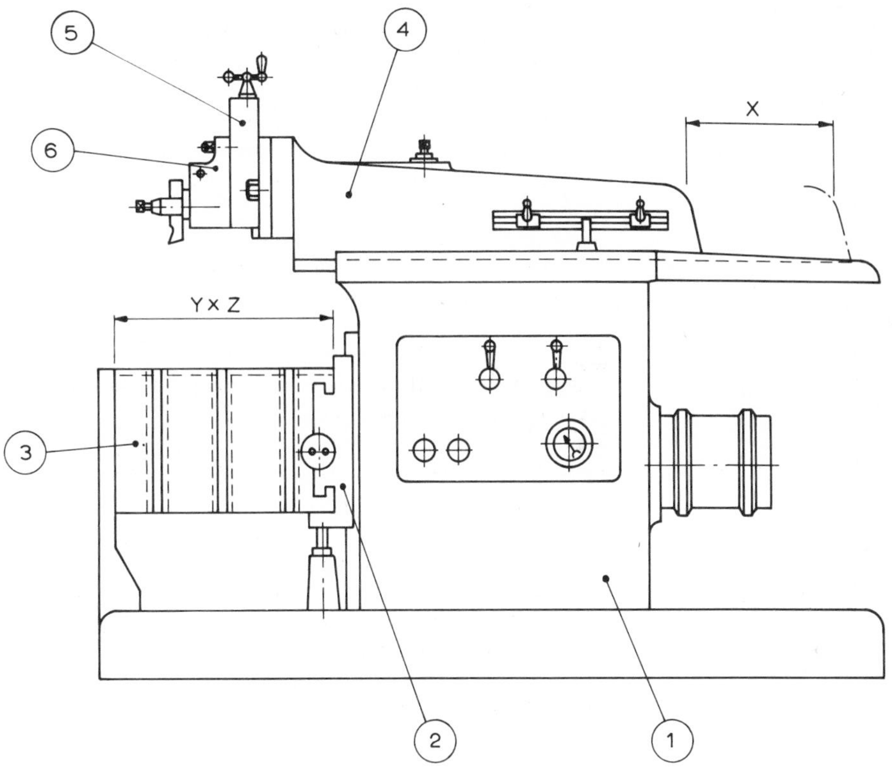

Fig. 1.19 Shaping machine

Dim.	Feature
Z	Table width
Y	Table length
X	Stroke

Item no.	Description
6	Tool-box
5	Head-slide
4	Ram
3	Table
2	Main slide
1	Body

15

3) TABLE The table is machined all over, and is provided with tee-slots on both the top and sides. It can be traversed across the main slide by either mechanical or hydraulic power. This cross-feed of the table is expressed in mm/cut.

4) RAM The ram reciprocates along slideways on top of the column, the forward stroke being the cutting stroke and the reverse stroke being idle. On mechanical machines, the motion of the ram is provided by means of a crank mechanism which incorporates a slotted-link quick-return system. Hydraulic mechanisms also provide a quick return stroke, the ram speed (and hence the tool cutting speed) being controlled by a throttle valve. As can be seen from fig. 1.19, the stroke of the ram is adjustable for length. It has the tool-head mounted at the front.

5) HEAD-SLIDE This may be fed vertically up or down the slideways, which themselves may be set at any desired angle (usually up to 60° either side of the vertical).

6) TOOL-BOX (see section 2.4) The tool-box incorporates the tool-post (into which the tool is clamped) and also the clapper-box, which is hinged so that it allows the tool to lift over the work surface on the return stroke. The whole tool-box may be set to an angular position when machining vertical and angular faces, thus giving natural clearance to the tool as it swings clear of the work on the return stroke.

Safety features
Many of the built-in safety features are similar to those used upon other machine tools, but, in addition, reciprocating machines have their own problems. Adequate clearance must always be allowed between a reciprocating slide in its maximum outwards positions and any obstruction such as a wall or machine. As shown in the diagram, ram guards are fitted to shaping machines at the rear, extending beyond the extreme back position of the ram.

It is difficult to visualise an efficient guard which could be used to cover the machining zone without seriously obstructing the operator; however, chip guards may be fitted at the front of the table to prevent hot metal chips flying dangerously.

Standard operations
Certain standard settings are used to produce the desired workpiece profile, some of which are shown in fig. 1.20.

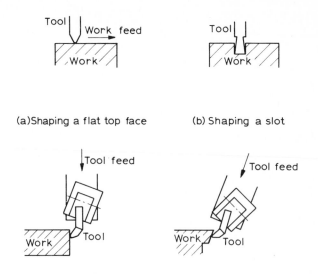

(a) Shaping a flat top face (b) Shaping a slot

(c) Shaping a vertical face (d) Shaping an angular face

Fig. 1.20 Shaping operations

Depending upon its size and shape, the work may be clamped directly to the top or side of the work-table or held in a machine vice. Exceptionally, a dividing head may be used where indexing of the workpiece is required.

1.7 Production of work on a drilling machine
A *drilling machine* is one of the simpler machine tools to understand, and is in general use throughout the engineering industry. The average drilling machine has a single vertical spindle, but production machines may be multi-spindle and of either vertical or horizontal configuration. Drilling machines are available to suit any size or class of work, and the main operations which may be carried out upon them are drilling, countersinking, counterboring, reaming, spotfacing, and tapping.

The pedestal machine may be regarded as the basic type, this being shown in fig. 1.21, together with its close relations the bench sensitive drilling machine and the radial drilling machine.

Figure 1.21 shows a vertical pedestal or column machine with the clean simple lines that are a feature of all modern machine tools. The most important dimensions–*X*, *Y*, and *Z*–have been added, these giving some indication of the size

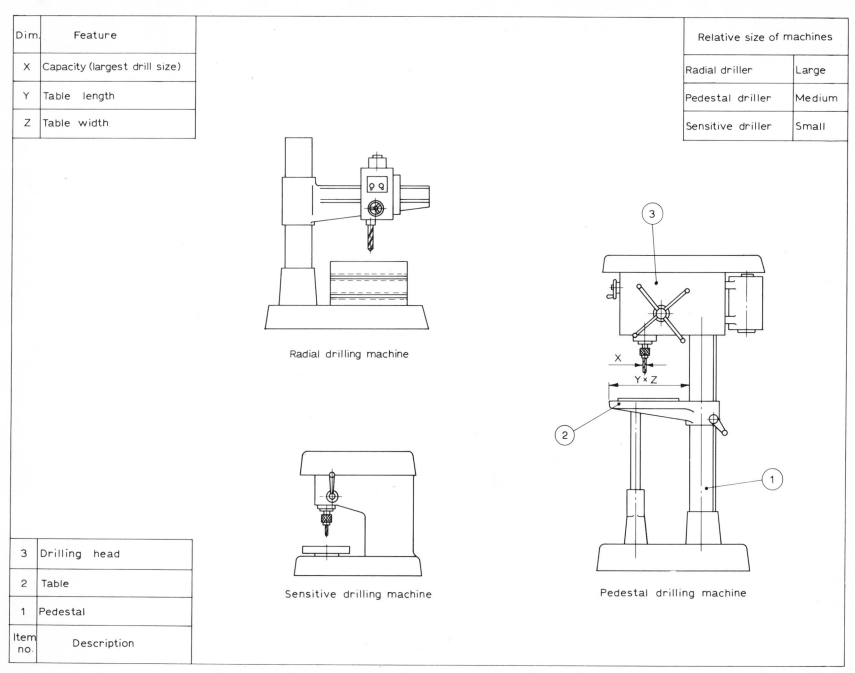

Dim.	Feature
X	Capacity (largest drill size)
Y	Table length
Z	Table width

Relative size of machines	
Radial driller	Large
Pedestal driller	Medium
Sensitive driller	Small

Radial drilling machine

Sensitive drilling machine

Pedestal drilling machine

3	Drilling head
2	Table
1	Pedestal
Item no.	Description

Fig. 1.21 Drilling machine

of job the machine is capable of handling. Drilling machines are usually referred to in terms of capacity (X dimension), this being the maximum diameter of hole that can be drilled in mild steel. This also then gives an indication of the maximum spindle speed available. The capacity of most pedestal machines is of the order of 50 to 100mm. The diagram also shows a bench-type sensitive machine and a radial-arm machine. The sensitive machine is intended for light work, having a capacity up to 12mm and utilising a sensitive hand feed to the drill spindle. The radial machine is usually of massive proportions, having the drilling head carried upon a radial arm which may be swung about the column axis. It is therefore intended primarily for use upon large workpieces, although it will be found to be a versatile machine.

The major features are as follows.

1) PEDESTAL This is the main column mounted on the base and, as drawn, is a round slideway up or down which the table may be moved.

2) TABLE The table may have a square working surface with tee-slots, as drawn, or it may be round and pivoted to turn into any desired position. Additional rigidity is provided by using the telescope jack under the table.

3) DRILLING HEAD The drilling head is carried on the top of the pedestal and contains the drill spindle and the means of driving it, viz. through gears or, alternatively, through a vee-belt drive which may give a stepped range or an infinitely variable choice of spindle speeds. All drilling machines have the one common design feature of a rotating spindle carried in an outer sleeve which may be fed vertically by a rack-and-pinion drive.

The spindle (or 'quill', as it is sometimes called) has a taper bore into which can be fitted drills, chucks, or tapping attachments (see section 2.4).

Safety features

The drilling process tends to look innocuous, but of course a rotating drill or cutter must be treated with respect, and should therefore always be guarded. This particularly prevents hair/head or finger injuries. Work should always be clamped, either in a vice, a jig (fixture), or directly to the table, and should never be held in the fingers.

In addition to the normal safety features, some drilling machines are fitted with a slipping-clutch mechanism between the spindle and the drive mechanism. This stops overload on drills, thus preventing breakages and possible damage to operator or machine.

Standard operations

These are shown in fig. 1.22. All the operations from (b) to (f) require an existing hole, which is followed by a second operation as drawn. Some of the tools have a pilot which guides the cutting edges in the correct relationship to the existing hole. Operations (b), (c), and (d) are carried out in order that screw or bolt heads may be accommodated in the hole.

Contrast fig. 1.22 with fig. 1.15, where the tool is stationary and the work rotates.

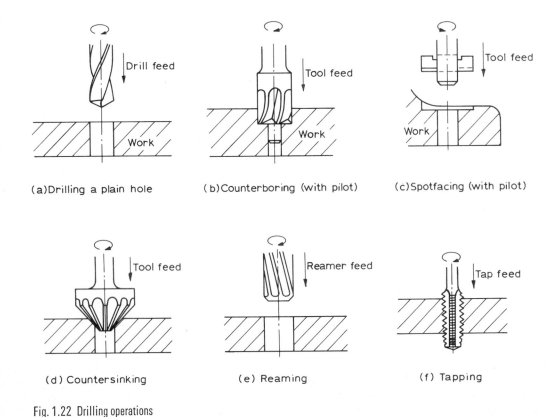

(a) Drilling a plain hole   (b) Counterboring (with pilot)   (c) Spotfacing (with pilot)

(d) Countersinking   (e) Reaming   (f) Tapping

Fig. 1.22 Drilling operations

Finally, it may be said that the *choice of machine* and cutting tool to manufacture a particular component will depend upon many factors, including size of work; number required; material, accuracy, and finish specified; cost and time available; etc. The *sequence of operations*, or order in which the work is processed, must also be determined before manufacture commences, and we will show an example of this in section 2.5.

Exercise 1

1 A quantity of the guide bushes shown in fig. 1.23 was machined to the final dimensions on the external features, using a centre lathe with standard cutting tools and equipment. To ensure concentricity of all diameters, and squareness of all faces to the bore, the bushes were held on a plain mandrel located in the previously finished bore.

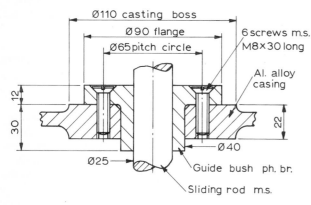

Fig. 1.23 Guide-bush assembly

a) With the aid of sketches, explain why the term 'generating' may be applied to the turning and facing operations.

b) If the 40mm diameter portion were found to be tapered after turning, enumerate the faults in the alignment of the lathe elements which would cause this error. What would be the possible effect on the geometry of the faced portion?

Use simple line diagrams to illustrate part (b) of the answer.   EMEU

2 Show how the lathe can be used for the following processes: (a) generating, (b) forming.   EMEU

3 Describe the action and purpose of a clapper-box on a shaping machine.   UEI

4 Sketch a centre lathe and indicate on your sketch the following: (a) the distance between centres, (b) the swing, (c) the length of bed.   UEI

5 a) The 18mm diameter hole is to be produced in the small casting shown in fig. 1.24. The face marked A is also to be machined flat and square to the hole. The operations may be carried out on either a lathe or a drilling machine.

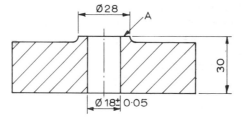

Fig. 1.24

For each case, neglect the method of holding, but show by means of simple line diagrams all the stages involved. Indicate clearly the relative movements of the tool and work.

b) For each case, show *one* error in the geometry which would result from incorrect procedure or misalignment of tool and work.   NCTEC

6 a) Show, using simple diagrams, the stages and tool-settings required to produce the profile shown on the component in fig. 1.25 using a shaping machine. It may be assumed that the outside dimensions of the block are already completed.

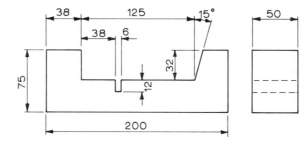

Fig. 1.25

b) What precautions are necessary to ensure that the form is produced at right angles to the face of the block?   NCTEC

7 a) Sketch a horizontal milling machine and indicate on your sketch the following: (i) the base, (ii) the column, (iii) the drawback bolt, (iv) the knee, (v) the overarm, (vi) the arbor support.

b) Give *four* safety precautions which should be observed when operating a milling machine. NCTEC

8 By means of diagrams, demonstrate the geometric shapes which can be produced on a lathe, a shaping machine, a milling machine, and a drilling machine, using basic machine-tool movements.

9 a) Explain the difference between a surface contour which is formed and one which is generated, and give two examples of each type.

b) Show by means of diagrams with short explanatory notes the two methods of producing a tapered surface of cylindrical section using a centre lathe. Indicate the limitations of each of these methods.

10 a) Sketch a horizontal milling machine and label the main features.

b) Make a list of the built-in features on the machine which contribute to operator safety.

c) Sketch, in third-angle projection, views in three planes of a suitable cutter-guard for this machine. Show the method used for fitting it to the machine and for adjusting it for capacity.

2.1 The basic metal-cutting wedge

A *single-point cutting tool* may be considered as a simple *wedge*, as shown in fig. 2.1.

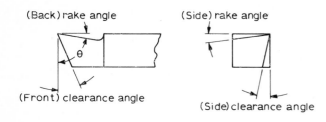

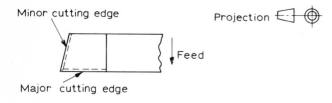

Fig. 2.1 Fundamental cutting wedge

The tool shown would be classed as *right hand*, in that it would be fed from right to left across the work. As drawn, with angle θ less than 90°, the tool has a *positive rake angle*; with angle θ greater than 90° it would be called a *negative-rake* tool. Rake and clearance angles ground upon a square section tool-bit give the *wedge-shaped* form to the tool, which enables it to remove metal from a workpiece by a *shearing* action. This shearing action creates bad frictional conditions between chip and tool, leading to high stresses and temperature being imposed upon the tool cutting face. Waste metal is removed from the workpiece in the form of either *discontinuous* or *continuous chips*, these conditions being shown in fig. 2.2, where a plain turning operation has been chosen to illustrate the shearing action of metal machining.

Chips of the type shown in fig. 2.2(a) are associated with:

a) brittle workpiece material,
b) low cutting speed,
c) small rake angle,
d) large chip thickness.

Chips of the type shown in fig. 2.2(b) are associated with:

a) ductile workpiece material,
b) high cutting speed,
c) large rake angle,
d) small chip thickness.

A cutting lubricant (and coolant) applied at the cutting zone can reduce the friction and improve tool life and work finish. Failure to use an efficient *cutting fluid*, in the case of continuous-chip cutting conditions, can lead to the formation of a *built-up edge* on the tool. This condition gives poor work finish, and eventually leads to the failure of the tool cutting edge.

RAKE ANGLE It can be seen from fig. 2.2 that the face over which the chip passes is inclined at an angle called the *rake angle*. This is varied for different work materials–in general, the more ductile the material, the greater the rake. As rake increases, cutting forces and hence the power required decrease. It should be noted that the tool will still cut even if the rake angle is zero or negative. Typical rake angles for various workpiece materials are shown below.

Work material	Rake angle
Nickle–chrome alloy steel	−5° to 8°
Cast iron	0° to 3°
Medium-carbon steel	3° to 10°
Mild steel	20° to 25°
Hard brass or bronze	0° to 3½°
Aluminium alloys	10° to 15°
Copper	15° to 30°

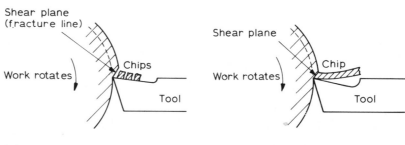

(a) Discontinuous chip   (b) Continuous chip

Fig. 2.2 Chip formation

The reader will notice that a range of values, rather than one value only, is given for each workpiece material. This is because:

a) each material shown can be obtained in a range having a variety of properties–for example, there is a series of commercial nickel–chrome alloy steels available, having varying degrees of hardness, toughness, etc.;

b) each material shown can be obtained in a range having a variety of forms–for example, brass can be obtained in the form of bar, plate, casting (sand- or die-), forging, or hot pressing; steel may have a clean bright finish prior to machining, or may have a scaled black finish–this influences the choice of rake angle;

c) the rake value is influenced a little by the type of cutting-tool material used. This may be:

   i) high-speed steel (average composition 0·7% carbon, 18% tungsten, 4% chromium, 1% vanadium),
   ii) stellite (average composition 33% chromium, 20% tungsten, 3% carbon, 44% cobalt),
   iii) cemented carbide (such as tungsten carbide clamped or brazed to a carbon-steel shank),
   iv) ceramic (average composition 85% aluminium oxide, plus other oxides or carbides–used in 'throw-away' tip form, clamped in a special holder),
   v) diamond (used as small tips held in a special holder for fine machining).

CLEARANCE ANGLE This angle is required to prevent the cutting tool from rubbing against the workpiece. A tool will not cut efficiently without adequate *clearance*. This value is affected by the geometry of the process, as is illustrated in fig. 2.3, which shows the average condition at (b)

compared to the two extreme conditions. Figure 2.3(c) illustrates the situation where a secondary clearance angle may be required in addition to the primary angle.

Cutting-tool nomenclature

Differing names are sometimes applied to the basic wedge cutting angles, particularly in other countries, though much progress has been made in establishing a single standard terminology. The nomenclature (system of naming) used BS 1296, 'Single-point cutting tools': part 2, 'Nomenclature', should be used. This specification is based upon the principle of categorising a cutting tool according to the *normal rake system*.

Essentially, this system defines the rake and clearance angles in a plane which is perpendicular, or normal, to the cutting edge. As the geometry of a single-point tool varies from point to point along the cutting edge (see fig. 2.1), it is necessary to select a point on the tool cutting edge where one wishes to position the reference plane. (The reader is now probably beginning to appreciate that the deceptively simple-looking single-point tool has hidden complications!)

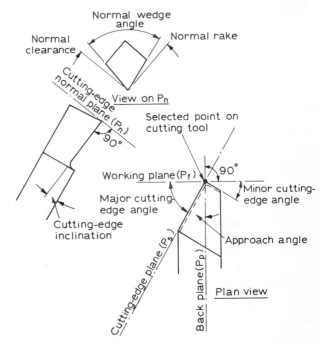

Fig. 2.4 Cutting-tool angles

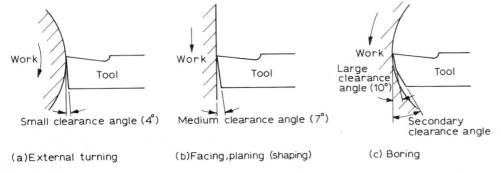

(a) External turning (b) Facing, planing (shaping) (c) Boring

Fig. 2.3 Tool clearance angle

Figure 2.4 shows the important planes and angles referred to in a normal rake system. If this diagram is compared with fig. 2.1, it will be noticed that the tool shape has been altered, since the major cutting edge is now shown inclined at an angle of less than 90° to the working plane. This has the general effect of reducing the power required for metal removal, and gives an increased life to the cutting edge.

2.2 The wedge principle in standard cutting tools

Single-point cutting tools, as used for turning or shaping, are in effect simple wedges, and the rake and clearance angles are easily identified, as in fig. 2.4. With the rotary cutting tools used for milling and drilling, however, these angles are not so obvious, and the typical wedge cutting shape is not always easy to identify. In order to show the application of the *wedge principle* to cutting tools, we will first consider milling cutters.

(a) End-milling cutter

(b) Peripheral-milling cutter

Fig. 2.5 Wedge principle in milling cutters

Figure 2.5 shows an end-milling cutter and a peripheral-cutter tooth, with the wedge cutting shape shaded. The end-milling cutter has positive rake, and the peripheral cutter has zero rake.

Figure 2.6 shows a twist drill and a reamer with the wedge shape shaded. The rake angle of the drill is given by the helix angle of the flute, which is positive as drawn. The reamer is shown with zero rake.

(a) Twist drill

(b) Reamer

Fig. 2.6 Wedge principle in drilling tools

In the case of cutters which cut on both the periphery and the end face (an end mill, for example), the application of the wedge principle can be observed in the side view and end view.

2.3 Drawings of standard cutting tools
Tools for lathes and shaping machines

As stated in section 2.1, BS 1296 should be referred to for the nomenclature of single-point cutting tools. Figure 2.7 shows the plan profiles of some of the more common lathe or shaper tools.

The tools shown in fig. 2.7 have the following uses:

(a) – general roughing or finish-surfacing work (lathe or shaper);
(b) and (c) – surfacing up to a corner or facing (lathe or shaper);
(d) – general facing work (lathe or shaper);
(e) – parting off (lathe) or grooving (lathe or shaper);
(f) – screw-cutting vee-threads (lathe);
(g) – boring holes (lathe).

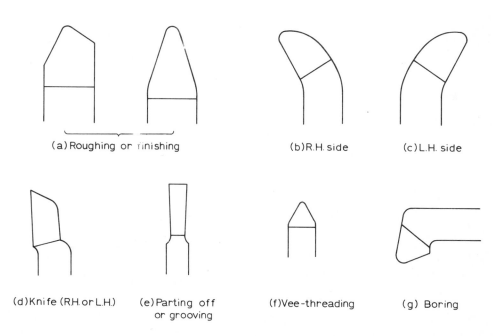

(a) Roughing or finishing

(b) R.H. side

(c) L.H. side

(d) Knife (R.H. or L.H.)

(e) Parting off or grooving

(f) Vee-threading

(g) Boring

Fig. 2.7 Plan profile of tools

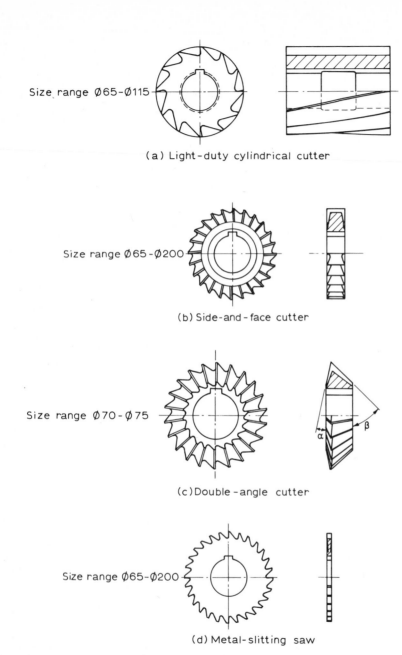

Size range Ø65-Ø115

(a) Light-duty cylindrical cutter

Size range Ø65-Ø200

(b) Side-and-face cutter

Size range Ø70-Ø75

(c) Double-angle cutter

Size range Ø65-Ø200

(d) Metal-slitting saw

Fig. 2.8 Milling cutters for arbor mounting

Tools for milling machines

BS 122: part 1 covers the range of standard milling cutters for use on horizontal and vertical machines. Figure 2.8 shows a selection of cutters which would be used on a horizontal machine, mounted upon the arbor.

The cutters shown have the following uses:

a) general flat surfacing work (high-power cutters can be obtained);
b) flat horizontal and/or vertical surfacing work;
c) angular surfacing, chamfering, grooving, or fluting–angles α and β are unequal as drawn; they may also be equal-angled or single-angled, in the latter case angle α being zero;
d) sawing or cutting narrow slots or grooves.

Figure 2.9 shows two typical cutters which would be used on a vertical machine by mounting in the spindle nose. They have the following uses:

a) general flat surfacing work and grooving,
b) production of tee-slots and undercuts.

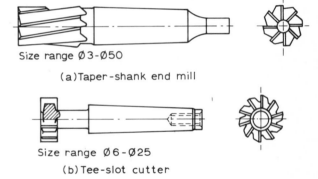

Size range Ø3-Ø50

(a) Taper-shank end mill

Size range Ø6-Ø25
(b) Tee-slot cutter

Fig. 2.9 Taper-shank milling cutters

End mills are available in a small size range (3–40mm diameter) having parallel shanks which can be held in a chuck; and a larger size range of 30–105mm diameter. The latter are called shell-end mills, and are fixed onto the end of a stub arbor which has a taper shank for fitting into the machine spindle nose.

The formal drawings of cutters shown in figs 2.8 and 2.9 should be compared with the informal diagrams of cutters shown in figs 1.10, 1.17, and 1.18. A fully-dimensioned formal working drawing is required for the manufacture of a cutter, but an informal diagram is sufficient to represent a cutter for the purposes of general illustration.

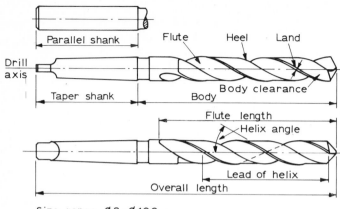

Size range Ø3–Ø100

(a) Twist drill

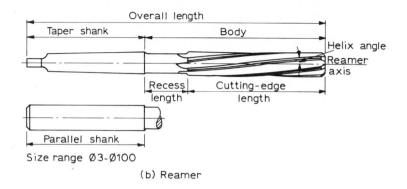

Size range Ø3–Ø100

(b) Reamer

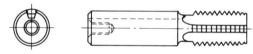

Straight flute shown – may also be helical flute

Size range Ø6×1.0 pitch – Ø60×5.5 pitch

(c) Machine tap

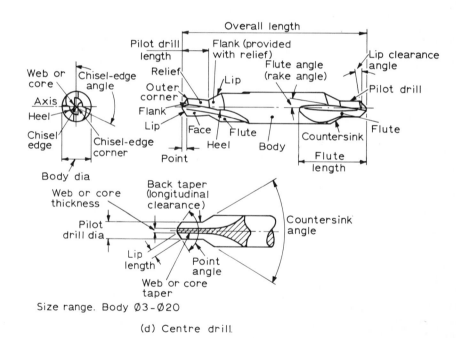

Size range. Body Ø3–Ø20

(d) Centre drill

Fig. 2.10 Internal cutting tools

Tools for drilling machines

BS 122: part 2, 'Reamers, countersinks, and counterbores', BS 328: parts 1 and 2, 'Twist drills, combined drills, and countersinks', and BS 949: parts 1 and 2, 'Screwing taps', cover the range of circular cutting tools used for hole production on drilling machines–and lathes too, of course. Figure 2.10 shows a selection of tools taken from these standards.

The tools shown in fig. 2.10 have the following uses;
a) general hole production,
b) finishing of holes to a high degree of accuracy,
c) internal threading,
d) combined drilling and countersinking.

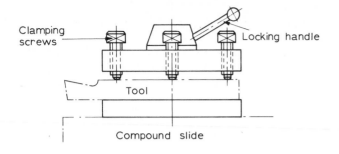

Clamping screws

Locking handle

Tool

Compound slide

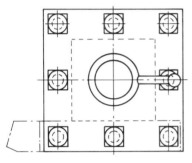

Fig. 2.11 Square tool-post

The centre drill in fig. 2.10(d) is shown with a complete nomenclature of features such as cutting angles, etc. This illustrates the fact that even such a simple cutting tool as a centre drill has a complex specification. Again, the drawings in fig. 2.10 might be compared with the diagrams in figs 1.15 and 1.22.

2.4 Tool location and holding devices

There is a variety of methods used for *positioning* and *clamping* various tools in various machines. We shall consider one method for each of the machines introduced in Chapter 1.

Centre lathe

Up to four cutting tools can be mounted in the *square tool-post* shown in fig. 2.11, which saves time compared with the alternative single-type tool-post. Each tool can be swung into position by rotating the tool-post and locking by means of the handle. Before being clamped by the screws, tools must be set to the correct height by using packing.

Shaping machine

The *tool-box* shown in fig. 2.12 accommodates only one tool, which is clamped in position in the tool-post slot by means of a screw. The hinged clapper-box allows the tool

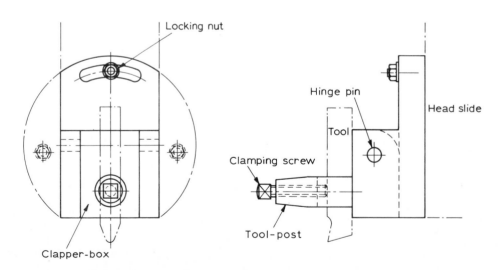

Locking nut

Hinge pin

Head slide

Tool

Clamping screw

Tool-post

Clapper-box

Fig. 2.12 Shaper tool-box

to lift freely over the work surface on the return stroke. The whole tool-box can be swung to a limited angular position on the head-slide and be locked in position with a nut.

Milling machine

The *arbor* shown fits into the horizontal-milling-machine spindle nose on a taper, being firmly locked with a draw-bolt. 'Dogs' fitting into slots on the arbor collar provide a positive drive. The cutter locates on the arbor, being positioned by the spacing collars, which pinch the cutter and clamp it under the influence of the locking nut. A driving key may be used on larger arbors. A bearing collar at the free end of the arbor rotates in the support-bracket bearing bush. Several cutters in a gang may be clamped on the arbor in any desired position.

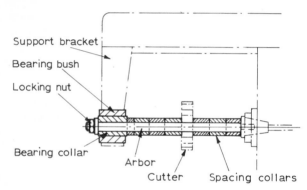

Fig. 2.13 Milling cutter on arbor

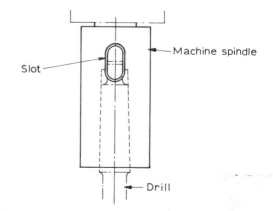

Fig. 2.14 Drilling-machine spindle

Drilling machine

A drilling-machine spindle has a 'sticking' taper bore (see BS 1660, 'Machine tapers': part 1, 'Self-holding tapers') which retains and drives the taper shank of the drill, reamer, or chuck in a frictional grip. A tang on the end of the shank fits into a slot, from which it can be extracted by means of a drift.

(Note that the larger spindle of a vertical milling machine has a non-stick taper, like the horizontal machine, and therefore requires a draw-bolt and dogs to retain and drive the cutting tool.)

2.5 Sequence of machining operations

As we stated at the end of Chapter 1, the *sequence of operations*, or order in which a component(s) is machined, and the type of machines and tools used depend upon many factors. It is obvious that the method used to produce a single component in a small tool-room will be quite different from that used to manufacture many components on a production line. Again, as stated earlier, the drawing may be regarded as the specification, and must be adhered to in all respects. The tool-maker, being a highly skilled craftsman, may carry the production *method* in his head, his skill and experience being incomparable aids. On the other hand, work to be produced upon the production line requires the sequence of operations to be formally set down on a *planning sheet*. This sheet will be drawn up in the production-planning department, and can be regarded as part of the complete specification.

The name and form of planning sheets varies from firm to firm, but certain essential information must be contained in them, viz. sequence of operations, machine and tools used, speeds and feeds, and sometimes the standard time to complete manufacture (at this stage we will ignore time estimating). In some instances the component is shown partly completed at each operation, thus indicating the changes in its shape as it is progressively machined to its final form.

To finish this chapter we will give an example of a completed planning sheet in fig. 2.16, which shows a sequence of operations for the manufacture of the part shown in fig. 2.15.

Assume a small batch of the components is required and the only machine tools available are a centre lathe, a milling machine or shaping machine, and a single-spindle drilling machine. Only standard tools and work-holders will

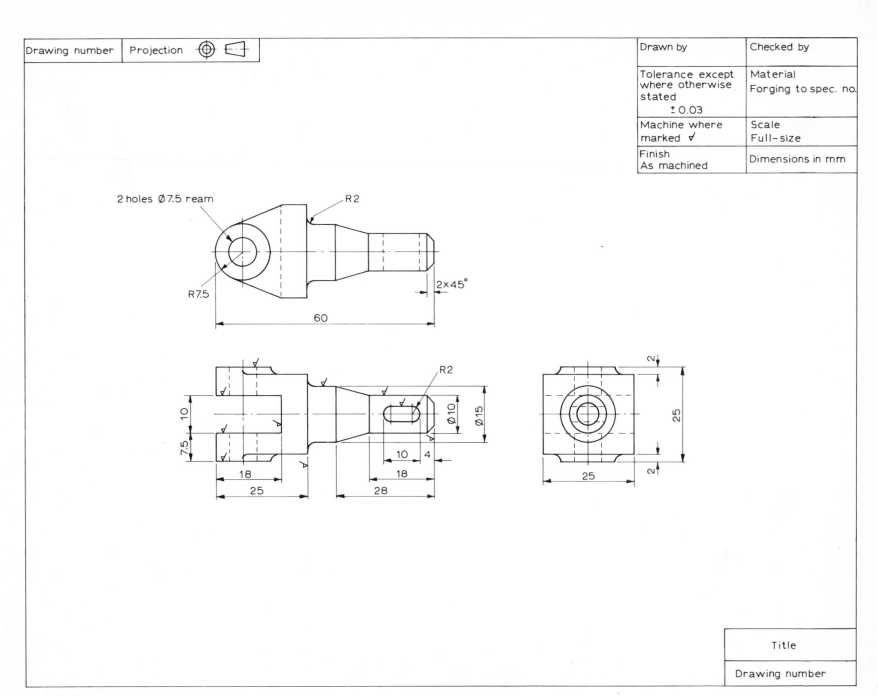

Drawing number	Projection ⊕ ⊏⊐		Drawn by	Checked by
			Tolerance except where otherwise stated ± 0.03	Material Forging to spec. no.
			Machine where marked √	Scale Full-size
			Finish As machined	Dimensions in mm

2 holes Ø7.5 ream

R2

R7.5

2×45°

60

R2

Ø10

Ø15

10

7.5

18

25

10   4

18

28

2

25

25

2

		Title
		Drawing number

Fig. 2.15 Component drawing

28

be used, the use of special tools and fixtures not being considered an economic proposition.

In practice, once cutter or *work speeds* and *feeds* have been estimated, the actual speeds and feeds chosen are those which are available at the machine.

It must be emphasised once again that there are several ways in which the component shown in fig. 2.15 could be machined down to size, depending upon the quantity required and the equipment available. The planning sheet in fig. 2.16 shows one possible sequence of operations; the 'best' method is that which gives the specified quality (degree of excellence) with regard to size, geometry, and finish at the *minimum cost*.

Exercise 2

1 Show, using simple outline sketches and brief notes, the stages and tools required to produce the component in fig. 2.17 using a centre lathe. All diameters must be concentric.   NCTEC

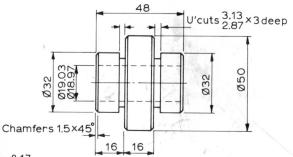

Fig. 2.17

2 a) State the purpose of rake and clearance angles on single-point cutting tools.

b) Make a sketch of a parting-off tool for use on a centre lathe, and indicate clearly on the sketch (i) the rake angle, (ii) the front clearance, (iii) the side clearance.

c) Show by means of simple sketches the differences between the cutters used on a milling machine for end milling and side-and-face milling.   NCTEC

3 a) Specify and describe using explanatory sketches *three* types of cutters for use with a plain horizontal milling machine.

b) Give *one* typical application where the cutters described in (a) would be used.   NCTEC

OP'N NO.	OPERATION	MACHINE USED	TOOLS USED (H.S.S.)	SPEED (rev/min)	FEED
①	Chuck and true up.	Centre lathe	4-jaw chuck	—	—
②	Rough turn 10–15 diameter.	Centre lathe	Round-nose rougher	500	0.2 mm/rev
③	Face end and chamfer.	Centre lathe	R.H. knife	800	Hand
④	Finish turn diameters, radius and shoulder.	Centre lathe	R.H. side	800	0.1 mm/rev
⑤	Clamp in machine vice.	Horizontal milling m/c	Machine vice	—	—
⑥	Mill side faces and 10 slot.	Horizontal milling m/c	10 wide x Ø90 side-and-face cutter	100	80 mm/min
⑦	Clamp in vice with vee support.	Vertical milling m/c	Machine vice Vee block	—	—
⑧	Mill slot.	Vertical milling m/c	End mill Chuck	1500	Hand
⑨	Clamp to machine table.	Pedestal drilling m/c	Clamp	—	—
⑩	Drill thro' 2 holes Ø7.	Pedestal drilling m/c	Ø7 twist drill	1000	0.15 mm/rev
⑪	Ream thro' holes Ø7.5	Pedestal drilling m/c	Ø7.5 machine reamer Chuck	100	0.5 mm/rev
⑫	De-burr.	By hand	File Scraper	—	—

Fig. 2.16 Planning sheet

4 a) Machine tools are used for either forming or generating surfaces. Give *one* example of each method, and explain how these methods of production are carried out.

b) Distinguish between the form of a lathe tool required for machining (i) a hard brittle material, (ii) a soft ductile material. NCTEC

5 a) Modern production lathes are fitted with a four-way square tool-post. Make a sketch of such a tool-post, showing clearly the methods used for (i) clamping the tool and (ii) setting the tool to its correct height.

b) Give *two* advantages the square tool-post has over a single-tool holder, and *one* advantage the single-tool holder has over the square tool-post. NCTEC

6 a) Show by means of clearly labelled diagrams how rake and clearance are put into (i) hacksaw blades, (ii) twist drills, (iii) helical slab mills.

b) With reference to the shaping machine, explain why (i) provision must be made for length adjustment of stroke, (ii) the ram position must be adjustable, (iii) the return stroke takes less time than the cutting stroke. NCTEC

7 a) Make a neat line diagram illustrating the basic metal-cutting wedge as applicable to metal-cutting tools.

b) Illustrate how the wedge functions in the following cutting tools, showing the position of the rake and clearance in each case: (i) a parting-off tool, (ii) a hacksaw, (iii) a twist drill.

c) State the purpose of clearance angles and explain why they should be kept to a minimum.

d) Sketch *two* examples of metal-cutting operations that require secondary clearance, giving the reason in each case. NCTEC

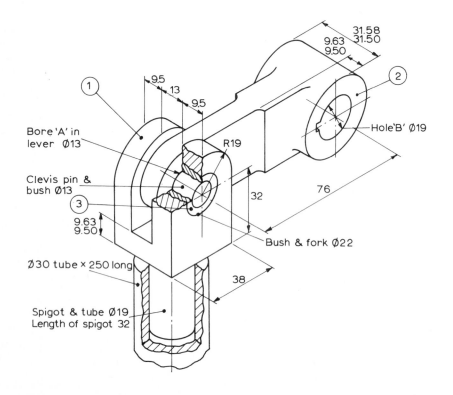

Fig. 2.18

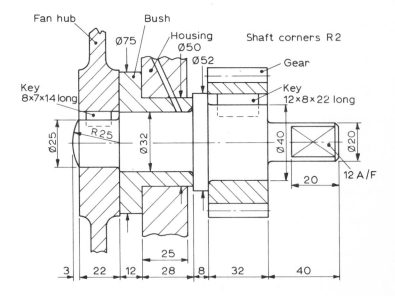

Fig. 2.19

8 Assuming the 19 mm diameter spigot in item 1, fig. 2.18, to have been previously turned, describe the operations necessary to mill the slot to the limits required, using a horizontal milling machine. Illustrate your answer with suitable sketches, paying particular attention to the setting method for the component and the type of cutter to be used. What conditions would be necessary for the accuracy of the slot when mounting the cutter on the arbor, and what precautions would you take for safe working during the cutting operation?   EMEU

9 a) Explain, with the aid of sketches, the term 'generating' when applied to the production of the 19 mm diameter spigot of item 1, fig. 2.18, by turning.
b) 'The term "forming" could be associated with "generating" when applied to the machining of the 19 mm radius on item 1, fig. 2.18.' With the aid of sketches, explain this statement.
c) Answer one of the following.
Either (i) Sketch a suitable tool for the operation in (a) above, assuming the material is forged mild steel. State a material which could be used for the tool, and give suitable values for rake and clearance angles.
Or (ii) Sketch a suitable cutter for the operation in (b) above, and show how rake and clearance is applied. State a suitable material which could be used for the cutter.   EMEU

10 The shaft shown in the assembly in fig. 2.19 is to be turned on a centre lathe, and flats are to be milled on a horizontal milling machine.
a) With the aid of sketches, explain how the shaft would be turned to ensure concentricity of all the diameters within close limits.
b) If the shaft is mild steel, state the cutting tools required and the material from which they would be made. Give this information in a suitable tabular form, to enable the tools to be drawn from a store.
c) Explain, with the aid of a sketch, the term 'straddle' when applied to the milling of the flats on the shaft.
EMEU

11 The medium-carbon steel forging shown in fig. 2.20 is to be produced in large numbers, and is to withstand considerable tensile stress.
Make out a complete sequence of operations for the machining of this component, giving details of cutting tools, small tools, and holding devices.   WJEC

12 a) Figure 2.21 shows a high-speed tool-bit of 12 mm × 20 mm section. Make neat orthographic sketches showing the end of this tool-bit suitably ground to take sliding cuts on mild-steel bar.
b) Make a neat sketch of a tool-holder suitable for holding a 12 mm × 12 mm high-speed tool-bit.   WJEC

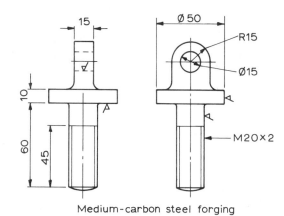

Medium-carbon steel forging

Fig. 2.20

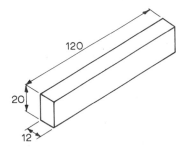

Fig. 2.21

Chapter 3
Dimensions, limits, & size control

1 Dimensions and tolerances

In section 1.1 the importance of adopting the correct procedure for showing *dimensions* and *tolerances* on a drawing was stressed. We will now examine this point in a little more detail. BS 308: 1972, 'Engineering Drawing Practice' should be referred to, so that standard practice is followed on all drawings. Part 2 of this standard lays down some general principles of dimensioning and tolerancing, the more important of which we will consider here.

Redundant dimensions

Each dimension necessary for the complete definition of a finished product should be given on the drawing, and should appear once only. Following from this, it can be said that no dimension should have to be deduced from other dimensions, nor should any dimensions have to be scaled directly from the drawing. Further, no more dimensions than are necessary to define the component should be shown; any superfluous dimension is called a *redundant dimension*.

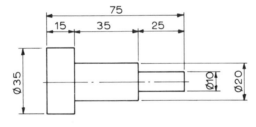

Fig. 3.1 Dimensioned component

Consider the dimensions of length shown in fig. 3.1. As the overall length is shown, one of the intermediate lengths is redundant and should not be given. This point becomes more important still when considering the accumulation of tolerances (see section 5.6).

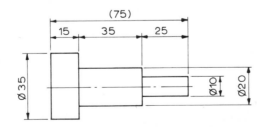

Fig. 3.2 Dimensioned component

Auxiliary dimensions

If it is thought that a redundant dimension could provide useful information, then it can be shown on the drawing as an *auxiliary dimension*.

Figure 3.2 shows the overall length given as an auxiliary dimension. It can be seen that this dimension is now included in parentheses (round brackets), and in effect has become a reference dimension. Auxiliary dimensions should not be toleranced, and do not govern acceptance of the product in any way.

Functional dimensions

These are dimensions which directly affect the functions of the product; nothing, therefore, could better illustrate the importance of taking care when applying dimensions to a working drawing. *Functional dimensions* should be expressed directly on a drawing in such a way that the chosen *datum features* (from which the dimensions are measured) are those which have a direct bearing on the function of the component in any assembly of which it may form a part. This is shown in fig. 3.3, which is extracted from BS 308.

This figure clearly shows the importance of the functional dimensions, which if incorrectly applied could adversely affect the function of the assembly. If any datum feature other than the one based on the function of the product is used, then finer tolerances are required than would otherwise be necessary. The following example, taken from BS 308 and shown in fig. 3.4, perfectly illustrates the principle.

Figure 3.4(a) shows the important dimension which must be correct to the specified limits of 12·00/11·92 mm after assembly. Figure 3.4(b) shows item 2 with the functional dimensions struck from the datum surface based on the function of the assembly. The assembly dimension shown at (a) is derived thus:

high limit $= 18·00 - 6·00 = 12·00$ mm
low limit $\ = 17·97 - 6·05 = 11·92$ mm

Figure 3.4(c) shows item 2 dimensioned from an alternative datum surface to give the same assembly dimensions, now derived thus:

high limit $= 18·00 - (12·00 - 6·00) = 12·00$ mm
low limit $\ = 17·97 - (12·03 - 5·98) = 11·92$ mm

It can be seen that a finer tolerance (0·03 mm) is required at (c) than the tolerance (0·05 mm) at (b). (The dimension

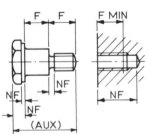

Fig. 3.3 Application of dimensions to suit design requirements
(Courtesy of British Standards Institution)
F = a functional dimension
NF = a non-functional dimension
AUX = an auxiliary dimension given, without tolerances, for information only

32

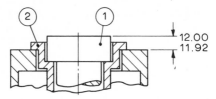

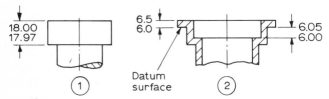

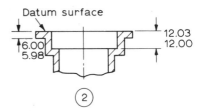

Fig. 3.4(a) Assembly drawing showing functional requirements

Fig. 3.4(b) Parts 1 and 2 dimensioned from functional datum surface

Fig. 3.4(c) Part 2 redimensioned using top surface as datum. Tolerances have had to be reduced to keep assembly within the limits 12·00 and 11·92

Fig. 3.4 Effect of changing datum surfaces
(Courtesy of British Standards Institution)

6·5/6·0mm does not affect the issue in situation (b), and need not be closely controlled.)

Tolerances

BS 308 states that *tolerances* should be specified for all dimensions affecting *functioning* or *interchangeability* (see section 3.2); in other words, all functional dimensions should bear a tolerance. Figure 3.4, for example, shows functional dimensions which are toleranced, and both the high limit and low limit of size are specified (see section 1.1). An acceptable alternative method of tolerancing is shown in fig. 3.5(b), where the tolerance is shown equally disposed about a basic size. Figure 3.5(a) shows the other method, for the purposes of comparison.

Similar principles of tolerancing may be adopted with angular dimensions, an example being shown in fig. 3.6.

In addition to tolerancing individual dimensions, a note can be applied to drawings specifying a general tolerance for a group of dimensions which are not critical. Such a note is shown in figs. 1.1, 1.6, and 2.15. The use of general notes greatly simplifies the preparation of drawings.

3.2 Limits and fits

Modern engineering manufacturing systems are based upon the principle of *interchangeability*; i.e. every one of a batch of mating parts is interchangeable with every other, without any fitting or modification being necessary. Interchangeable manufacture is possible only if two factors are established:
a) the type of *fit* which is required between mating parts, for example a push fit, a drive fit, etc.;
b) the *tolerance* (and hence the limits) which is to be allowed upon each dimension to give the specified fit.

The appropriate standard to be consulted in order to determine these two factors is BS 4500: 1969, 'ISO limits and fits'. Figure 3.7 shows a plain shaft which will enter the hole, giving a clearance fit. The allowable tolerances for both shaft and hole are indicated. The figure shows the following features.
a) High limit – the largest permissible dimension.
b) Low limit – the smallest permissible dimension.
c) Tolerance – the amount of variation which can be tolerated because of the inherent inaccuracies of the manufacturing process. It is the difference between the high limit and the low limit.
d) Minimum clearance – the difference between the hole low limit and the shaft high limit.
e) Maximum clearance – the difference between the hole high limit and the shaft low limit.

(a)

(b)

Fig. 3.5 Tolerancing linear dimensions

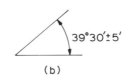

Fig. 3.6 Tolerancing angular dimensions

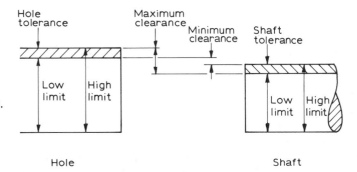

Fig. 3.7 Terminology of a limits and fits system

The limits, and hence the clearance between any shaft and hole, will be chosen by the designer to give the required functional characteristics for the assembly, i.e. to give the appropriate fit to suit the conditions under which the assembly must work. Fundamentally, there are three types of fit, these being the clearance fit, the interference fit, and the transition fit.

CLEARANCE FIT Figure 3.7 shows a clearance fit where the largest possible shaft is less in diameter than the smallest possible hole.

INTERFERENCE FIT This type of fit results from the smallest possible shaft being greater in diameter than the largest possible hole.

TRANSITION FIT The word 'transition' means 'in between two conditions', in this case, in between clearance and interference. With this fit, the smallest shaft could give a clearance fit and the largest shaft could give an interference fit. The transition fit is not in common use.

Note that running fits and push fits are types of clearance fits, and drive fits, press fits, and force fits are types of interference fits, all these terms being self-explanatory.

The *type of fit* is determined by the magnitude of the *tolerance* and the *fundamental deviation* (F.D.). The latter fixes the disposition of the tolerance zone (on shaft or hole) with respect to the nominal or basic size. The wide range of tolerances and fundamental deviations which can be selected from BS 4500 gives many types of fit.

BS 4500: 1969, 'ISO limits and fits'

This is a system of limits and fits which covers a size range up to 3150 mm. It has eighteen tolerance grades from which to choose, these being designated IT01, IT0, IT1, IT2, etc., up to IT16. The standard unit of tolerance used is 0·001 mm.

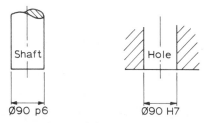

Fig. 3.8 Dimensioning a fit by symbols

The F.D. is designated by capital letters for holes, i.e. 'A' to 'ZC', and similarly by small letters for shafts, i.e. 'a' up to 'zc'. There are 41 F.D.'s of differing magnitude specified.

We will consider one example of an interference fit for a nominal size of 90 mm selected from the standard, using an H7 hole (i.e. of F.D. class H and tolerance grade 7) combined with a p6 shaft (i.e. of F.D. class p and tolerance grade 6). This fit gives limits of

H7 hole	p6 shaft
90·035 ⎫ mm diameter	90·093 ⎫ mm diameter
90·000 ⎭	90·071 ⎭

This fit gives a maximum interference of 0·093 mm and a minimum interference of 0·036 mm, and would be classed as a press fit, i.e. the shaft would have to be pressed into the hole at assembly. Figure 3.8 shows how the parts could be dimensioned on a drawing, using symbols instead of sizes for the limits. Only the nominal size is actually written as a dimension.

3.3 Surface finish

The principles underlying *surface-finish* measurement can be found in BS 1134, 'Centre-line average height method for the assessment of surface texture'. The production process used to bring a component to its final shape will leave irregularities on the surface as a result of the inherent action of that process. For example, machining on a lathe will result in traverse feed marks being present on the work surface; the finer the feed marks, the better the quality (degree of excellence) of the surface texture. As stated in section 1.1, there is a close relationship between the surface finish and the tolerance; the machine tool which is capable of producing a component size to a very fine tolerance will also produce a fine surface finish. Likewise, a component having 'open' dimensions (i.e. with no limits imposed on the size variation) is unlikely to require a high-quality surface texture. If an unnecessarily high surface finish is specified, as well as a tight tolerance, then money will be wasted in achieving it.

We are not concerned here with the methods used to measure surface texture, but rather how it should be specified upon a drawing. The surface roughness is designated in *micrometres* (μm), where one micrometre = 0·001 mm. A surface very carefully produced by milling or turning might have a surface roughness of 5μm. This value may be regarded as the average height and depth about a mean line passing through the surface profile, and is known as the CLA ('centre-line average') value. Below are given the ranges of surface-finish values (from best to worst) which may be expected from the four machining processes examined in Chapter 1.

Process	Range of surface texture (µm)
Drilling	20 to 100
Shaping	10 to 100
Milling	4 to 100
Turning	4 to 100

Where it is necessary to indicate a particular quality of surface texture, BS 308 recommends that this value should be included with the machining symbol of the type shown in fig. 1.6 and shown again here in fig. 3.9. This symbol may be applied to any line on the drawing representing a surface, as at (a) and (b), or may be used as a general note, as at (c).

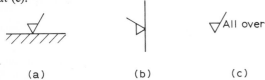

(a)   (b)   (c)

Fig. 3.9 Machining symbols

Two examples of surface-texture symbols are shown in fig. 3.10, a single value being represented on the left, and both maximum and minimum values being represented on the right. This indicates that the larger diameter surface must have a surface-texture value no greater than 20 µm, and the smaller diameter surface must have a finish lying between 5 and 10 µm.

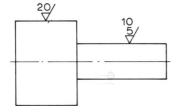

Fig. 3.10 Application of surface-texture symbols

3.4 Conventional representation
Standard parts, such as simple compression springs, may have to be drawn many times by draughtsmen; this leads to a great amount of time being wasted in drawing unnecessary

Title	Subject	Convention
External screw-threads (detail)		
Internal screw-threads (detail)		
Screw-threads (assembly)		
Thread inserts		
Splined shafts		
Serrated shafts		

Fig. 3.11 Conventional representation of common features (Courtesy of British Standards Institution)

detail. *Conventional representation* should be adopted in such cases. This means that a common engineering object may be shown in a drawing by means of a simple diagrammatic (schematic) representation. The accepted convention for such diagrams may be found in BS 308. A selection of conventional representations of common features is shown in figs 3.11 to 3.13; these examples are simple and self-explanatory.

Title	Subject	Convention
Straight knurling	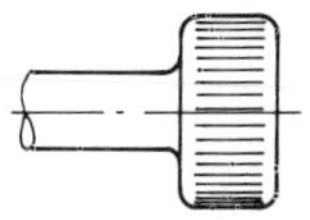	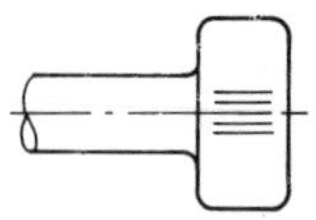
Diamond knurling	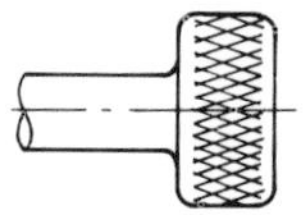	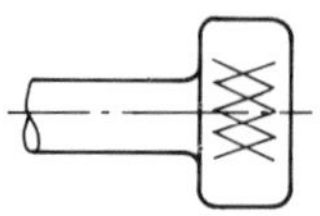
Square on shaft	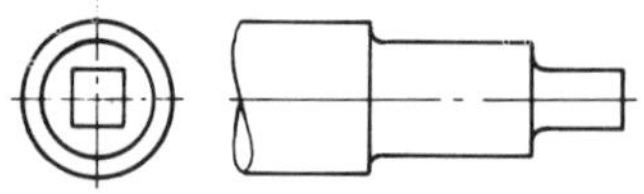	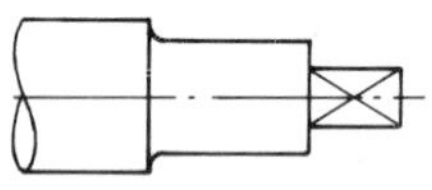
Holes on circular pitch	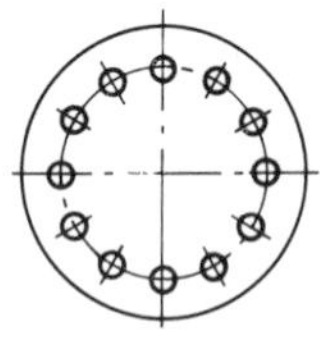	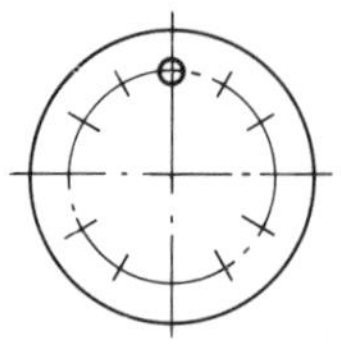
Holes on linear pitch	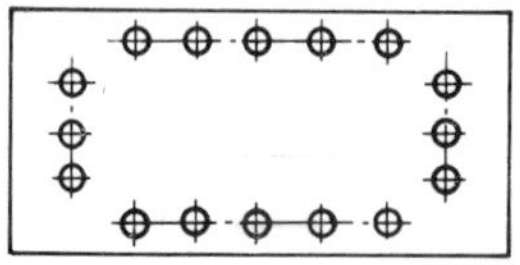	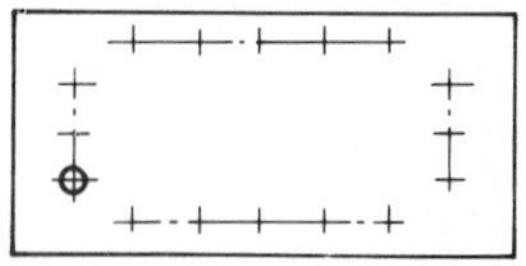
Bearings	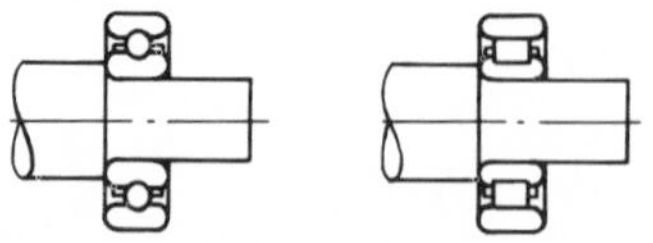	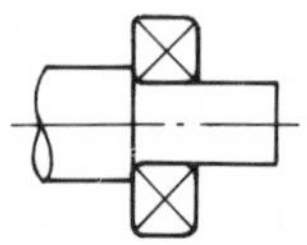

Fig. 3.12 Conventional representation of common features
(Courtesy of British Standards Institution)

Title	Subject	Convention
Interrupted views	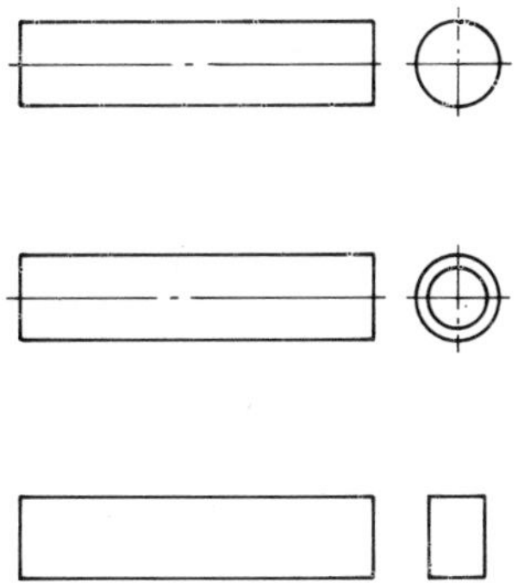	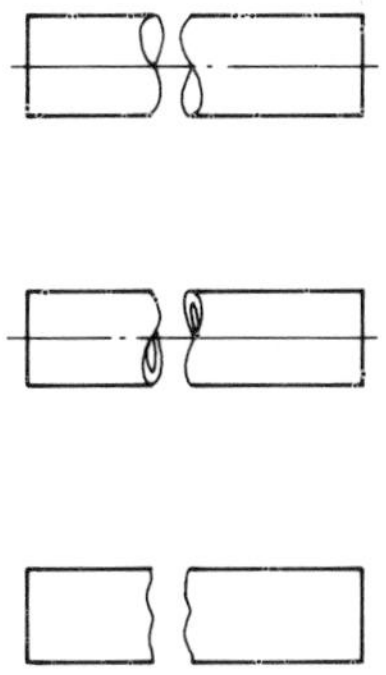
Repeated parts	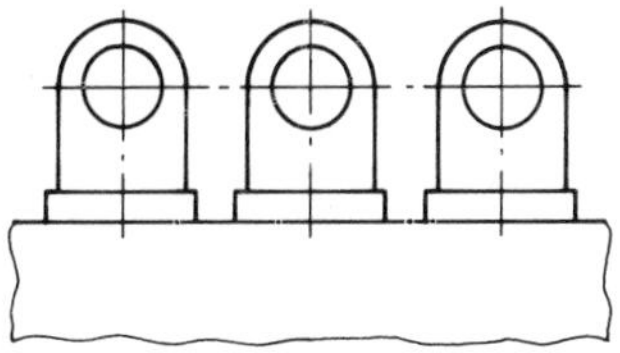	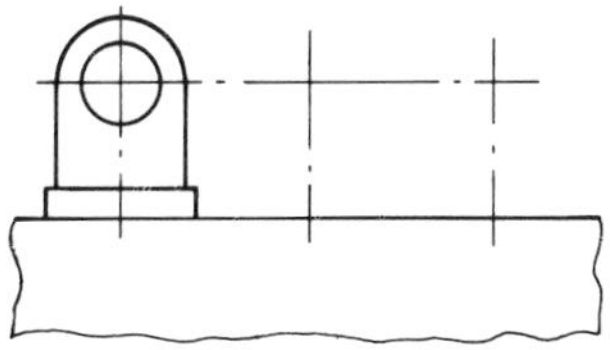
Semi-elliptic leaf springs	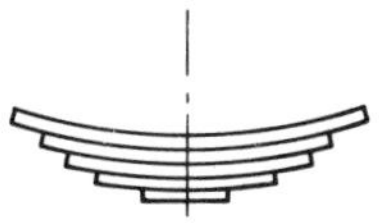	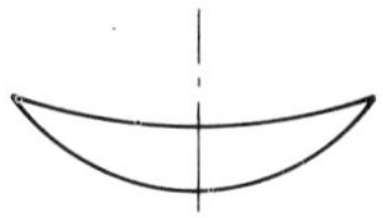
Semi-elliptic leaf spring with eyes	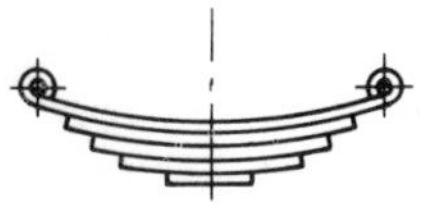	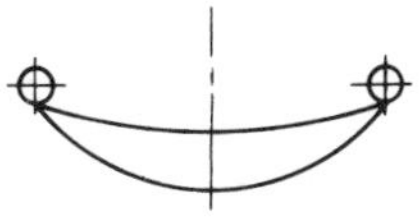

Fig. 3.13 Conventional representation of common features
(Courtesy of British Standards Institution)

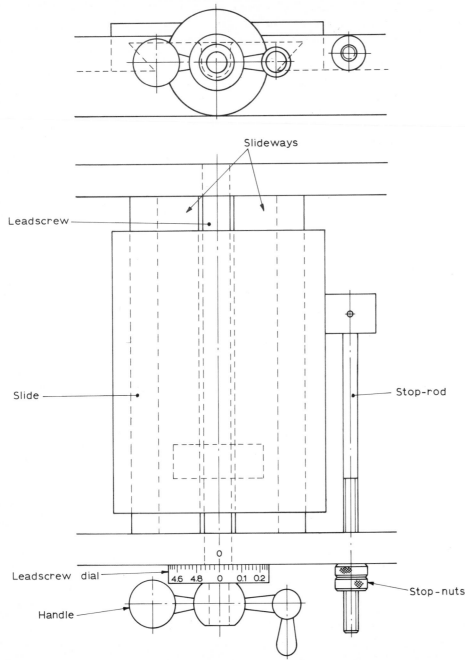

Slideways

Leadscrew

Slide

Stop-rod

Leadscrew dial

4.6 4.8 0 0.1 0.2

Handle

Stop-nuts

Fig. 3.14 Leadscrew-controlled slide

3.5 Control of size on machine tools

When a drawing shows a component dimension with limits added, it means that the dimension must be *controlled* within the *specified limits* during the course of manufacture. In the case of a machining process, it therefore follows that one must be capable of controlling the movement of the tool- (or work-) holding slide within the required limits. If the limits are close, then a high degree of accuracy in the movement of the slide is required. The slide moving in its slideways may carry the tool-holder and tool past the work; or alternatively a table may carry the work past a cutter. In either case the principle is the same, and we will now consider three well-established methods of precisely controlling the movement of a machine slide along a slideway.

a) *Control by leadscrew*

This is the time-honoured method of utilising a rotating *leadscrew* of known pitch to push the slide (attached to the leadscrew) along the desired amount. The leadscrew might be rotated mechanically through a gear-drive, or by hand. Figure 3.14 illustrates the principle of operation of such a slide, and it will be noticed that the leadscrew provides both the motive power and the means of controlling the slide movement.

When the leadscrew is rotated by means of the handle, the slide moves along dovetail slideways. A stop-rod may be provided, to enable the slide to be fed to the same pre-set position each time, without reference to the leadscrew dial. The *calibration* (units of size indicated by the graduations) of this dial is important. If the leadscrew pitch is 5 mm, say, then one complete revolution of the leadscrew will move the slide a distance of 5 mm. If 250 equal divisions are marked upon the dial, then a screw movement of 1 division will give $5/250 = 0.02$ mm movement of the slide.

The principle used here is *analogous* (similar in certain essential respects) to that of a micrometer (see section 3.6). The accuracy of the slide movement is obviously dependent upon the accuracy of the leadscrew pitch and of the dial graduations. Some sophisticated machines have devices fitted which make due correction for any leadscrew pitch error.

b) *Control by scale*

This method utilises a *graduated linear scale* to control the slide movement; hence the leadscrew would be used only to provide the motive power. Alternatively, some other means such as a pneumatic or hydraulic cylinder, might be used to move the slide. Figure 3.15 shows the principle.

Figure 3.15 shows the slide of fig. 3.14 with a precision scale fitted along one edge. The datum point on the slide shows its position in relation to the scale. The scale will be sensibly and accurately graduated to suit the size and accuracy of movement required. A very finely calibrated scale might require an optical device, such as a microscope, mounted above it to assist the naked eye. The degree of accuracy is further enhanced by the use of a vernier scale (see section 3.6).

c) Control by end standard

Again, as in method (b) above, a leadscrew or other means might be used to move the slide, and the means of controlling the distance moved by the slide is independent of the medium used for moving it. Figure 3.16 shows the principle.

Figure 3.16 shows the slide of fig. 3.14 with a DTI (dial test indicator) fitted to a bracket on the side of the slide.

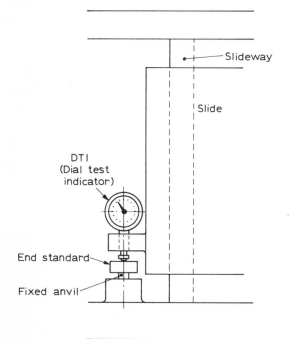

Fig. 3.16 End-standard-controlled slide

First the slide can be moved until the DTI anvil engages the fixed machine anvil. The DTI scale is now 'zeroed', thus establishing a datum, and hence the slide is set in its initial position.

An *end standard*, such as a slip gauge, is now obtained with length equal to the desired slide movement. After the slide has been moved, the end standard is placed between the anvils (as shown in fig. 3.16), and the slide position is finally adjusted until the DTI indicating pointer takes up the original zero position on the scale. The slide is then set in its final position, and has moved a distance equal to the length of the chosen end standard.

The principle behind this method is similar to that of comparative measurement (see section 3.6).

3.6 Measurement of work size

After completion of a workpiece to its final size and shape, it will be necessary to inspect it in order to ascertain whether it complies in all respects with the drawing specification. Sizes which bear limits can be checked by measurement or gauging. The former method makes use of instruments which reveal the workpiece size, and hence the extent of the error, if any. *Gauging* does not reveal the actual workpiece size, but only indicates whether or not a dimension is within the specified limits. We shall examine the principles of gauging in section 3.7.

The instruments we will discuss in this section are all capable of *measuring* dimensions to a degree of accuracy suitable for the machine tools covered in Chapter 1. The best degree of accuracy which, with care, can be achieved by these machining processes is given below:

Process	Tolerance (mm)
Turning	0·02
Milling	0·025
Shaping	0·05
Drilling	0·05

We will now consider three well-established measuring methods which correspond in principle with the three methods of slide-movement measurement discussed in section 3.5.

a) Direct measurement by micrometer

A *micrometer* makes use of a precision screw, the angular rotation of which leads to accurate linear displacement of the spindle. The micrometer is attached to a caliper frame

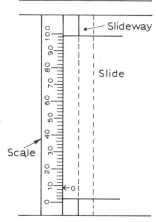

Fig. 3.15 Scale-controlled slide

for the purposes of external measurement. Figure 3.17 is a diagram of a caliper or outside micrometer.

The bore of the barrel is precisely screwed with a thread of 0·5mm pitch. The hardened spindle and the thimble are attached to each other and screw through the barrel bore. Therefore, one turn of the thimble causes the spindle to move a linear distance of 0·5mm from the hardened anvil. The barrel is graduated in equal increments of 0·5mm, and the end of the thimble is divided around its circumference into 50 equal divisions. If the thimble is turned through one of these divisions, then the spindle will move 1/50 of a pitch = 0·5/50 = 0·01mm.

Work size is measured between the end faces of the anvil and spindle. The distance they are apart can be read directly by observing the number of whole turns uncovered on the barrel (in units of 0·5mm) and adding on the part of a turn as indicated on the thimble (in units of 0·01mm). Figure 3.18 shows two metric-micrometer readings in a diagrammatic form.

Some micrometers are fitted with a locking nut (for locking the spindle at a pre-set distance from the anvil) and a ratchet stop (for screwing the spindle up to the work at uniform pressure or 'feel'). The size range of micrometers is 25mm, i.e. 0–25, 25–50, 50–75mm, etc., and internal and depth versions are available.

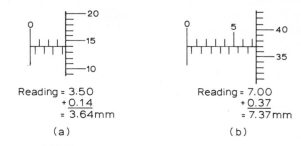

Reading = 3.50
+0.14
= 3.64mm

(a)

Reading = 7.00
+0.37
= 7.37mm

(b)

Fig. 3.18 Metric micrometer readings

The *accuracy* (the agreement of the instrument reading and the actual size) with which measurements can be made using a micrometer depends largely upon the skill of the user and the degree of precision with which the instrument is made. However, the *resolution* (the smallest unit of measurement to which the instrument can be read) of a micrometer can be improved by means of a vernier scale on the barrel. This enables readings to be made in units of 0·002mm. Let us now consider the principle of vernier scales.

b) Direct measurement by vernier

The accuracy with which a line scale (such as an engineer's rule) can be read depends, as with the micrometer, upon the skill of the user and the degree of precision to which the scale has been divided. If the scale is divided too finely (i.e. if the resolution is too small), difficulties arise in reading it, because the lines appear to blend together. This difficulty can be overcome by a *vernier scale* used in conjunction with the main scale. A vernier caliper is used for end measurement, either external or internal, the movement of the sliding jaw being indicated by the scales. This instrument is illustrated in fig. 3.19.

The sliding jaw is positioned approximately by hand. Its final precise position is achieved by means of the knurled fine-adjustment nut, before the jaw is finally locked. The main scale is divided into 1mm units, which are again sub-divided into equal 0·5mm units. The length of the vernier scale is 12mm, which is graduated in 25 equal divisions, the length of each of these divisions being 12/25 = 0·48mm. Therefore the difference in length between a division on the main scale and a division on the vernier scale is 0·5 − 0·48 = 0·02mm; this value represents the resolution of the instrument.

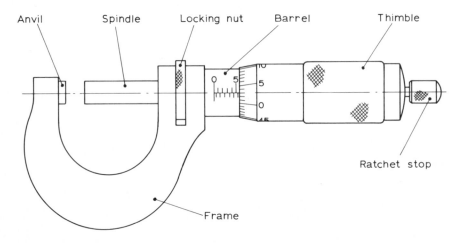

Fig. 3.17 Caliper micrometer

Anvil   Spindle   Locking nut   Barrel   Thimble

Ratchet stop

Frame

External work size is measured in between the flat inside faces of the jaws. The distance they are apart can be read directly by observing the main-scale reading up to the zero graduation on the sliding scale, then adding to it the vernier-scale reading. For internal work, the thickness of the two jaw ends must also be added. Figure 3.20 shows two metric-vernier readings.

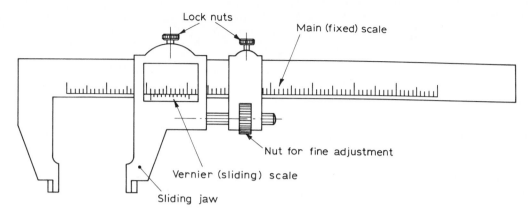

Fig. 3.19 Vernier caliper

Reading = 4.50 (Main scale)
+ 0.22 (11×0.02 Vernier scale)
= 4.72 mm

(a)

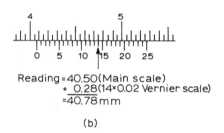

Reading = 40.50 (Main scale)
+ 0.28 (14×0.02 Vernier scale)
= 40.78 mm

(b)

Fig. 3.20 Metric-vernier readings

Another commonly used instrument utilising the vernier principle is the vernier height-gauge.

c) Comparative method of measurement
We will consider here the simplest version of this method. This requires an *end standard*, such as a *slip gauge*, against which the component dimension being checked can be compared. The principle involved is common to all measurement, viz. the *comparison* of an unknown value against a known value. A DTI (dial test indicator–a dial indicator or clock gauge) is also required, to record any deviation between the work and the standard. The set-up is shown in fig. 3.21.

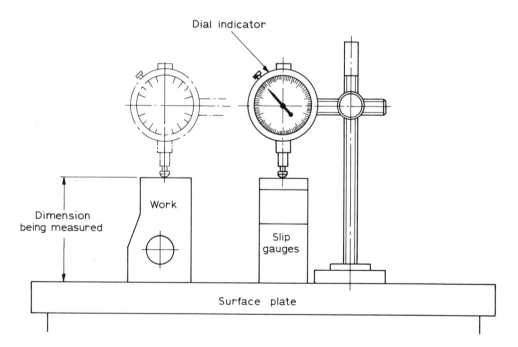

Fig. 3.21 Measurement by comparison

41

There are two variations in the method illustrated in fig. 3.21, as follows.

i) Build a pile of slip gauges up to the required height by '*wringing*' them together. ('Wringing' is the act of carefully joining gauges together by sliding and rotating them against each other such that tight contact is achieved.) In this case, the required height will be the specified dimension on the drawing (either the high limit or the low limit). Position the DTI until its anvil engages the top face of the slip-gauge pile, then 'zero' the DTI scale; hence a datum has been established. Now transfer the DTI to the work, and any deviation of the pointer from its original zero setting will be recorded in units of 0·001 mm. Therefore any difference between the standard size and the work size will be known, and it can easily be determined whether or not the work size is within the specified limits.

ii) In this alternative case, the slip-gauge height can be built up to the same size as the workpiece dimension by trial and error, using the DTI only as a *fiducial indicator* ('fiducial' means 'having trust or confidence'). When the pointer indicates the same zero reading for both work and standard, they must be of the same size.

Note that in (i) the difference (if any) between two sizes is determined, and in (ii) the actual size of the work is determined. With the *comparative method*, these alternatives are always available. The use of end standards ensures a greater accuracy than is possible using a line standard such as a rule. A typical set of slip gauges has 78 pieces comprising:

49 pieces in steps of 0·01 mm from 1·01 to 1·49 mm
19 pieces in steps of 0·5 mm from 0·5 to 9·5 mm
 7 pieces of 10, 20, 30, 40, 50, 75, and 100 mm respectively
 3 pieces of 1·0025, 1·005, and 1·0075 mm respectively
___
78 total

Sources of error in measurement

Before leaving this section, the reader's attention is drawn to the more important sources of error which can occur when using measuring equipment of the type we have just considered. It is possible that one or more of the sources considered below is present during a measuring operation.

a) MEASURING-TEMPERATURE There are two different ways in which *temperature* plays a part in causing errors of measurement:

i) there may be a difference in temperature between the workpiece and the standard – work fresh from a machining operation requires time to cool to the ambient (surrounding) temperature;

ii) there may be a difference in the materials (and hence the coefficients of linear expansion) from which the work and standard are made, in which case sizes should be compared only at the standard temperature of 20°C.

b) MEASURING-FORCE When using a caliper instrument, for example of the micrometer or vernier type, the accuracy of any measurements made will be affected by the magnitude of the *force* used to close the anvils or jaws across the work. The sense of touch, which indicates to the user that the instrument and the work are just in contact, with minimum pressure between them, is known as 'feel'. Much practice is required before one acquires this sensitivity of touch which leads to accurate and consistent results.

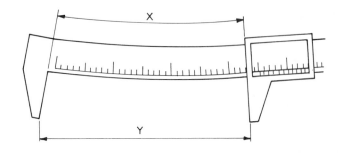

Fig. 3.22 Force effect of vernier caliper

Figure 3.22 shows in exaggerated form how a vernier caliper can be deflected because excessive force has been used in applying it across the work. If the jaws are forced over the work, causing curvature of the instrument as shown, then the length *X* which is registered on the scale is less than the true length *Y* between the jaws.

When using a ratchet-operated micrometer caliper, the problem of the operator's feel is removed. Note also that the same problem is not a factor when using a DTI in the comparative method of measurement.

c) HUMAN FAILINGS Certain measuring errors can occur which are the direct result of lack of skill on the part of the operator. Three important examples are as follows:

i) *Misalignment* of instrument with work (leading to what is known as *cosine error*); see fig. 3.23(a). Dimension *M*, which is the distance at which the micrometer anvils are set apart, will be registered on the scale instead of dimension *D*, which is the actual diameter of the work.

ii) *Parallax*. This can be defined as the apparent change in the position of an object due to a change in the position of the observer, and is an optical effect; see fig. 3.23(b). The object in this case is a micrometer scale, and the operator's eye is in the wrong position for an accurate reading. The situation is improved if '*x*' is kept to an absolute minimum.

iii) *Vernier acuity* (keenness or sharpness of observation). This is the ability of an operator to determine when two lines on adjacent scales are exactly in line with each other. Many people are unable to accomplish this with precision.

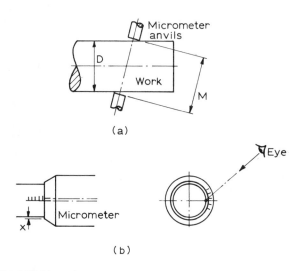

Fig. 3.23 Measuring-errors

3.7 Gauging of work

Limit gauges, as they are called, may be used instead of measuring instruments to check whether or not work is within the specified limits. Gauging, unlike measuring, cannot reveal the magnitude of any error which may be present; however, gauging is preferred for quantity-production work because:

a) it is simpler and quicker than direct measurement, requiring little skill, and is therefore cheaper;

b) solid fixed-size gauges are relatively easy to design and make, and are often manufactured in a firm's tool-room.

The disadvantages of gauging are:

a) as already stated, it reveals only whether work is right or wrong;

b) gauges are subject to manufacturing error and to errors caused by wear in use.

As an introduction to the subject, it will be sufficient to consider standard fixed-size gauges used for the checking of external diameters and internal diameters. Adjustable gauges are also used in industry, being suitable for a range of sizes.

Gauging of external diameters

A *caliper* or gap-type *gauge*, either single- or double-ended, is used for this purpose. 'Gauge-plate' (precision ground, flat, chrome-alloy steel) is commonly used for gauge manufacture, being hardened and tempered to resist wear.

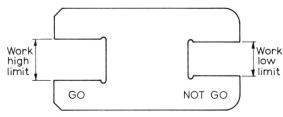

Fig. 3.24 Caliper gauge

Figure 3.24 shows a double-ended caliper gauge, often called a *snap gauge*. The work is acceptable if the GO end of the gauge passes over the work and the NOT GO end does not.

Gauging of internal diameters

A simple *plug gauge*, either single- or double-ended (as illustrated in fig. 3.25) is used for this purpose. The gauging members are hardened, tempered, and ground.

The gauge shown in fig. 3.25 would be suitable for checking moderate size holes; large holes would require a gauge of lighter design. Gauging members are often made separately from the handles, for reasons of economy. With plug gauges, the GO end is made to the work low limit and the NOT GO end to the work high limit, which of course is the opposite of a caliper gauge. Hence the gauge GO end should enter the hole of an acceptable component, and the shorter NOT GO end should not.

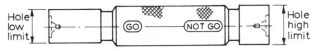

Fig. 3.25 Plug gauge

Exercises 3

1 a) Make neat sketches of limit gauges suitable for checking the 15 mm diameter hole and the 20 mm external diameter on the component shown in fig. 2.20.
b) Describe how both the hole and external diameter can be measured directly to within 0·01 mm.   WJEC

2 The drawing in fig. 3.26 is an example of incorrect dimensioning. Redraw the component full-size, and dimension the drawing using the correct dimensioning procedure according to BS 308.   NCTEC

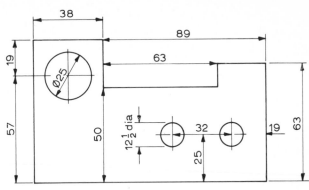

Fig. 3.26

3 Using the appropriate section of BS 4500, prepare a fully dimensioned drawing of a cast-iron flat-belt pulley to the following specification:
   outside diameter 250 mm,
   width of belt face 50 mm,
   bore diameter to a close running fit (H7/g6) on a
   55 mm shaft,
   boss diameter 80 mm,
   overall width of boss 70 mm.
   Assume any dimensions not given.   NCTEC

4 a) By means of diagrams, explain the meaning of the following terms: (i) tolerance, (ii) clearance fit, (iii) interference fit, (iv) transition fit.
b) What is 'interchangeability', and how is it obtained?
   UEI

5 Figure 1.23 shows a guide-bush arrangement. Using the data given in table 3.1, make a full-size working drawing, including all relevant dimensions and machining instructions, as follows: (a) a complete single-part drawing of the bush, (b) a 'scrap' detail for the machining of the bore and the tapped holes in the aluminium-alloy casing.
   The 25 mm diameter is an H7/f7 fit, and the 40 mm diameter is an H7/s6 fit.   EMEU

6 a) Sketch a typical plug gauge suitable for inspecting a bore 30·20/30·35 mm diameter. Dimension your sketch with suitable tolerances for the manufacture of the gauging diameters.
b) Write brief notes to show the essential differences between a gauge and a measuring instrument.   NCTEC

7 The casting shown in fig. 3.27 requires to be machined in the bore and on both outer faces. The three holes are to be drilled and spotfaced.

Nominal or basic size		Approximate designation of fit								
Over	Up to and including	Loose running		Precision running		Average location		Press (ferrous)		Press (non-ferrous)	
		H9	d10	H7	f7	H7	g6	H7	p6	H7	s6
mm 10	mm 18	+43 0	−50 −120	+18 0	−16 −34	+18 0	−6 −17	+18 0	+29 +18	+18 0	+39 +28
18	30	+52 0	−65 −149	+21 0	−20 −41	+21 0	−7 −20	+21 0	+35 +22	+21 0	+48 +35
30	40	+62 0	−80 −180	+25 0	−25 −50	+25 0	−9 −25	+25 0	+42 +26	+25 0	+59 +43

Table 3.1 Selection of fits from BS 4500:1969 (tolerances in units of 0·001 mm)

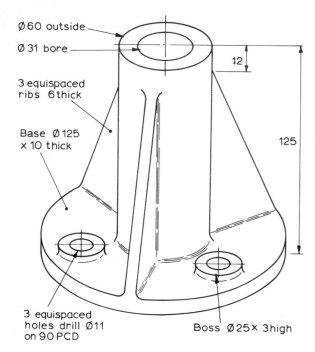

Ø60 outside

Ø31 bore

3 equispaced
ribs 6 thick

Base Ø125
x 10 thick

3 equispaced
holes drill Ø11
on 90 PCD

Boss Ø25× 3high

12

125

Fig. 3.27

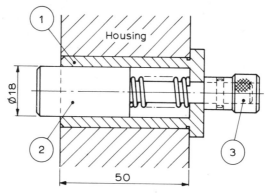

H7/g6 fit between items ① and ②

H7/p6 fit between item ① and housing

Housing

1

Ø18

2

50

3

Dimensions not given to be estimated

Fig. 3.28

Draw a sectional front elevation and a plan. Fully dimension the drawing, and add the necessary machining details in accordance with BS 308. NCTEC

8 a) Explain the principle and purpose of indexing dials as used on standard machine tools.

b) The cross-slide of a centre lathe has a leadscrew of 2 mm pitch. If there are 50 divisions on the indexing dial, calculate the depth of cut for an indexing of 12 divisions. WJEC

9 a) Compare measuring with gauging with regard to (i) the results and (ii) the circumstances under which each would be used.

b) Sketch a typical caliper gauge suitable for checking a shaft 35·00 ± 0·5 mm diameter.

10 Figure 3.28 shows a spring-loaded plunger. Make complete detailed working drawings of the parts labelled 1, 2, and 3, fully dimensioned in millimetres. Two views in orthographic projection are required for each part. Include on each drawing (a) a suitable general tolerance, (b) suitable toleranced dimensions for specific fits (ref. BS 4500), (c) suitable machining symbols.

11 a) Illustrate a reading of 6·82 mm upon a micrometer and vernier scale respectively.

b) Compare a caliper micrometer and vernier caliper, stating the advantages and limitations of each type of instrument.

c) Show what is meant by the 'comparative' method of measurement, using a dial test indicator as the comparator.

12 With the aid of sketches, outline the operating principles for achieving precise movement of a machine-tool slide by three different methods. List the advantages and limitations of each of the methods.

13 a) Discuss conventional representation as an aid to draughtsmanship. Do you think there are any disadvantages in using such a system?

b) Sketch the conventional representation for the following common engineering items: (i) an external screw-thread, (ii) a compression spring, (iii) a diamond-knurled nut, (iv) a single-track ball-bearing, (v) a plain spur gear.

Chapter 4
Marking out

1 Dimensioning hole centres

When specifying hole-centre dimensions on a drawing, in order to fix the position of holes, a draughtsman has the choice of two alternative methods. These are (a) angular spacing around a pitch circle and (b) rectangular coordinates. The choice of the dimensioning method used will depend upon the circumstances of the work, as we will shortly see, and yet again it will become apparent that the manner in which the work is dimensioned will determine the methods used to prepare and manufacture the work (see section 4.2). Let us consider the alternative dimensioning methods in turn.

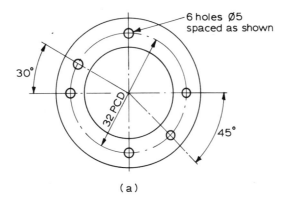

(a)

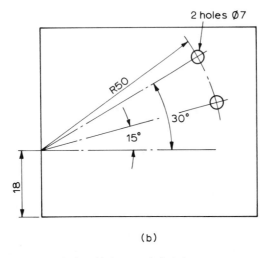

(b)

Fig. 4.1 Dimensioning of holes on a pitch circle

a) *Angular spacing around a pitch circle*

A *pitch* is the distance between successive hole centres, threads, or gear teeth. A *pitch circle* is a circle around which the holes are spaced, and its diameter is the *pitch-circle diameter* (PCD). Figure 4.1 shows examples of holes spaced (a) on a pitch circle and (b) on a part pitch circle.

Figure 4.1(a) shows a symmetrical circular component which lends itself to the chosen method of dimensioning, the pitch-circle centre being the work centre. The example shown at fig. 4.1(b) is not so clear-cut, and would lend itself more to the alternative method: rectangular coordinates.

b) *Rectangular coordinates*

An *ordinate* is one of two lines used to fix the position of a point, and a *rectangular coordinate* is one of a group of ordinates whose general pattern is rectangular. Figure 4.2 shows the component of fig. 4.1(b) now dimensioned in rectangular coordinates.

It will be seen from fig. 4.2 that the horizontal and vertical dimensions from the hole centres to two edges are now specified. These dimensions have been calculated thus:

$$48 \cdot 295 = 50 \cos 15°$$
$$43 \cdot 300 = 50 \cos 30°$$
$$43 \cdot 000 = (50 \sin 30°) + 18$$
$$30 \cdot 940 = (50 \sin 15°) + 18$$

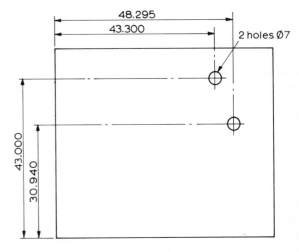

Fig. 4.2 Rectangular coordinate dimensions

In effect, the two edges from which the dimensions have been struck have become the datum (see section 1.1), instead of the centre (pole) of the part pitch circle. This will therefore determine the method used to mark out (or 'set out') the hole centres before drilling. Method (a) is sometimes referred to by the term 'polar coordinates'.

4.2 Marking out hole centres

If a large quantity of drilled components were required, then a locating device such as a drill jig would be used to position the holes at the correct centres on the work. We will assume here, however, that we are dealing with individual parts which are to have the hole centres positioned by marking out. The instruments used to mark out lines on work surfaces are scribers, odd-leg calipers, dividers, scribing blocks or surface gauges, and vernier height-gauges. A dot-punch is used to improve the clarity of a line. Angle plates, jacks, vee-blocks, etc. are used as accessories (additional items). However, it is not intended to describe the skills required for marking out, but rather to concentrate upon the principles involved; as an example, we will examine the way in which the hole centres are marked out for the part shown in fig. 4.1(b) and fig. 4.2. Remember that this is the same part dimensioned in two different ways, which means that it will be necessary to mark it out using a method which is in accord with the dimensioning method used on the drawing. Let us consider each way in turn.

Marking out the hole centres shown in fig. 4.1(b): the order of marking out will be as shown in fig. 4.3.

Marking out the hole centres shown in fig. 4.2: the order of marking out will be as shown in fig. 4.4.

One might reflect that the end result is the same in any event, in that the hole centres are positioned at the same points in each case. Does it matter then, in this case, how the draughtsman communicates his requirements to the craftsman who is engineering the job? The answer is 'yes', even in such a simple example as the one illustrated. If the draughtsman dimensions in polar coordinates when it is more convenient for the fitter to mark out in rectangular coordinates (or vice versa), then the fitter must carry out calculations (hence increasing the possibility of error) in order to complete his task. It cannot be stated too often that the chosen manufacturing method will reflect the original information specified on the drawing.

Before drilling the holes, greater accuracy of positioning

Opn no.	Sketch	Operation	Tools used
①	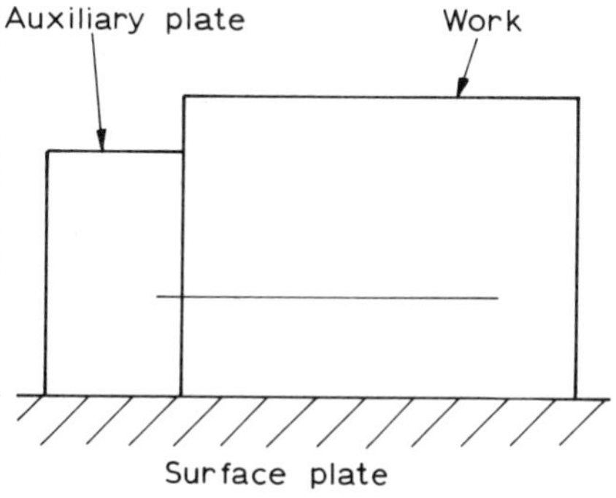 Auxiliary plate — Work / Surface plate / All work faces machined square	Set work on bottom datum edge. / Scribe horizontal line 18 from datum.	Scribing block
②	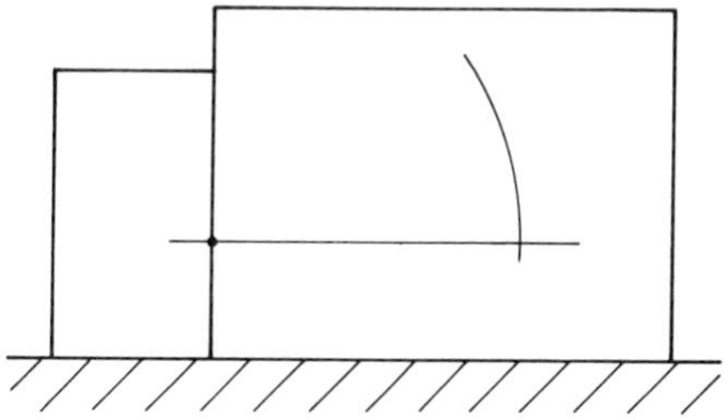	Centre-punch centre (pole). / Scribe part pitch circle of 50 radius.	Centre-punch / Dividers
③	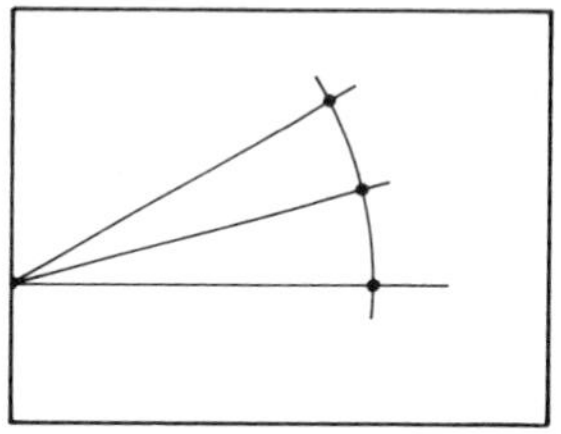	Scribe lines struck from pole at 15° & 30° respectively. / Centre-punch hole centres.	Scriber / Straight edge / Protractor / Centre-punch

Fig. 4.3 Marking out hole centres (polar coordinates)

can be achieved by scribing and dot-punching the hole circumference [as shown in fig. 4.5(a)], or by scribing and dot-punching an enclosing box [as shown in fig. 4.5(b)].

'Boxing' is the more accurate method, as clearer evidence of the original marking-out lines and the intended hole position is left after drilling.

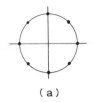 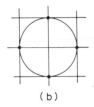

(a)   (b)

Fig. 4.5 Marking out holes

Opn no.	Sketch	Operation	Tools used
①	**Work** — All work faces machined square — Surface plate	Set work on bottom datum edge. Scribe horizontal line 30.940 from datum. Scribe horizontal line 43.000 from datum.	Vernier height-gauge
②		Set work on side datum face. Scribe horizontal line 43.300 from datum. Scribe horizontal line 48.295 from datum. Centre-punch hole centres.	Vernier height-gauge Centre-punch

Fig. 4.4 Marking out hole centres (rectangular coordinates)

4.3 Construction of profiles

When setting out the profile of a component on a drawing, the construction is first completed using thin continuous lines. The final outline of the profile, which is to be clearly visible, is then drawn in using thicker continuous lines. These two types of lines are indicated in fig. 4.6. Construction lines should be faint, and need not be rubbed out. Outlines should be clearly visible.

The profile shown in fig. 4.6 is made up simply of straight lines; their construction into a final outlined profile can easily be carried out using standard drawing instruments. Where a profile consists of a combination of straight lines and circular arcs, however, the construction may be more difficult. Some of the more common standard constructions are given below, and are illustrated in figs 4.7 to 4.11.

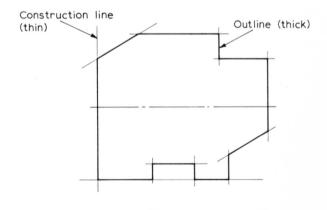

Construction line (thin)   Outline (thick)

Fig. 4.6 Outline of workpiece profile

a) Joining two straight lines at right angles by an arc of known radius

Draw two straight lines XX and YY normal to each other, to intersect at O (see fig. 4.7).

With centre O and radius equal to the required radius of the arc, R, draw arcs to intersect the straight lines at A and B.

With centres A and B, and the same radius R, draw arcs to intersect at O_1.

With centre O_1 and the same radius R, draw the required arc to join the straight lines.

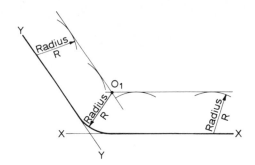

Fig. 4.8 Joining two straight lines by an arc

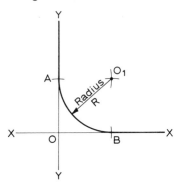

Fig. 4.7 Joining two straight lines by an arc

b) Joining two straight lines at an obtuse angle by an arc of known radius

Draw two straight lines XX and YY at the required angle to each other (see fig. 4.8).

At a distance from these lines equal to the required radius of arc, R, draw lines parallel to lines XX and YY to intersect at O_1.

With centre O_1 and the same radius R, draw the required arc to join the straight lines.

c) Joining two straight lines at an acute angle by an arc of known radius

The construction shown in fig. 4.9 is identical to that described above for fig. 4.8.

d) Joining an arc and a straight line by an arc of known radius

Draw the straight line XX and the circular arc YY, of radius R_1 struck from centre O, in their required positions (see fig. 4.10).

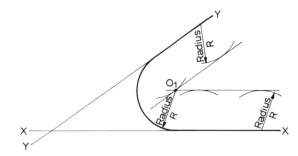

Fig. 4.9 Joining two straight lines by an arc

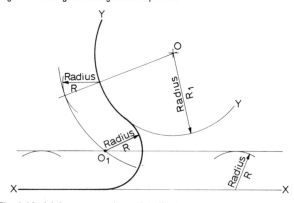

Fig. 4.10 Joining an arc and a straight line by an arc

At a distance equal to the required radius of arc, R, draw a line parallel to line XX.

With centre O and radius equal to $R_1 + R$, draw a second arc to intersect this straight line at O_1.

With centre O_1 and radius R, draw the required arc to join the original straight line and arc.

e) Joining two straight lines by two arcs of equal radius

Draw two straight lines XX and YY at the required distance apart (they need not necessarily be parallel).

Mark points A and B where they are required to be joined by arcs, and draw a straight line through these points.

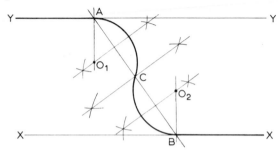

Fig. 4.11 Joining two straight lines by two arcs

Bisect A and B at C by drawing intersecting arcs (of greater radius than $\frac{1}{2}$AB) above and below C from centres A and B. Draw a straight line through the intersecting arcs.

Bisect AC and CB similarly.

Draw perpendiculars from A and B to intersect these bisectors at points O_1 and O_2.

With centre O_1 and radius equal to O_1A, draw an arc from A to C.

Likewise, with centre O_2, draw a similar arc from C to B to join the lines.

In addition to the examples given above, there are other similar constructions made up of combinations of straight lines and arcs which may be drawn using the same principles.

Figure 4.12 shows an example of a plate having external and internal profiles made up of straight lines and arcs which must blend smoothly with each other. The construction lines have been left in so that the drawing methods used are clear.

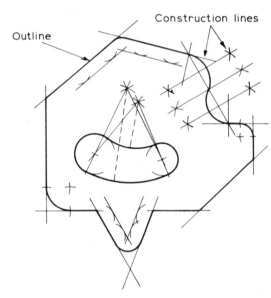

Fig. 4.12 Construction of profiles

4.4 Marking out profiles

Components may be marked out to assist the machining or hand-working of the part to its final shape or profile. This method is satisfactory where only a small number of parts is required. Metal will be removed up to the profile lines

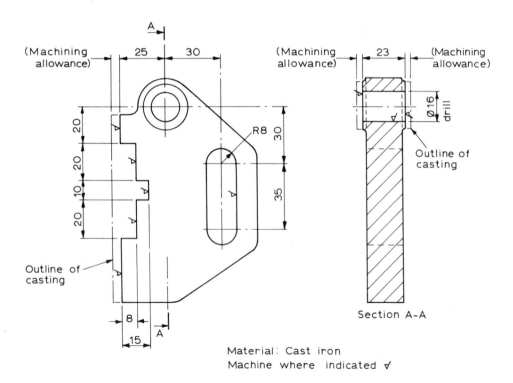

Material: Cast iron
Machine where indicated ∨

Fig. 4.13 Cast component

until the dot-punch marks are split. Where greater accuracy is required, the part must be brought to its final shape and size by measurement—the marking-out lines acting merely as a guide. Smooth faces to be marked out are coated with copper sulphate, which acts as a good background against which the lines can be seen. Likewise, rougher surfaces, such as cast metal, are coated with a white marking fluid.

As seen in section 4.2, two important principles are involved in marking out. These are:

a) selection of the correct marking-out method to suit the dimensions specified on the drawing,
b) selection of the correct datum point(s) or edge(s) from which all the other dimensions are struck.

In the case of a simple component like that shown in fig. 4.3, having straight square-machined edges before the hole centres are marked out, little difficulty should be experienced. However, in the case of a rough casting, for example, which is to have a surface profile marked out prior to machining, more difficulty will be encountered in selecting datum faces and setting up. Let us consider such a case now. Figure 4.13 shows a drawing of a casting which must be marked out ready for machining where indicated. Figure 4.14 shows the sequence in which the various lines are marked out.

It should be noted from fig. 4.13 that the primary form of the component is a solid iron casting, sufficient material being left on (as shown by the dotted outline) to allow machining to the final shape. The only dimensions shown are those necessary to mark out the profiles and centre lines.

One or two points should be noted from the method indicated in fig. 4.14.
a) The edge to be machined (the left-hand edge) is not chosen as the first datum, because, if the 16 mm hole centre were struck from the casting outline, it would finish up off-centre from the boss. Therefore, the boss centre is first found, and then used as the datum, hence ensuring concentricity of hole and boss. All other dimensions are struck from this datum. This procedure also ensures that the correct machining allowance is left on the left-hand face of the casting.
b) All drawing dimensions are given from the hole centre (not from an edge), hence the marking-out method is in accord with the drawing specification.
c) When marking out the 23 mm width, again the casting outline is not used as a datum face. Instead, the centre line

Opn no.	Sketch	Operation	Tools used
1		By scribing arcs, find centre (datum point) of boss.	Odd-leg calipers
2	Angle plate / Work / Surface plate	Set work with top face horizontal and clamp. Scribe (a) boss centre (datum). " (b) lines 10,17 and 25 up from datum. " (c) lines 22,30 and 38 down from datum.	Scribing block
3		Set work with datum line vertical and clamp. Scribe (a) boss centre (datum). " (b) lines 20,40,50 &70 down from datum. " (c) lines 30 and 65 down from datum. Punch Ø16 and R8 centres.	Try-square Scribing block Centre-punch
4		Scribe (a) Ø16 circle. " (b) 8 radii. Dot-punch all scribed outlines for clarity.	Dividers Dot-punch
5	Work / Parallel / Surface plate	By scribing arcs find centre point. Scribe (a) centre line (datum). " (b) line around boss 11.5 up from datum. " (c) line around boss 11.5 down from datum. Dot-punch outlines.	Odd-leg calipers Scribing block Dot-punch

Fig. 4.14 Marking out profiles

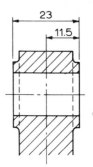

Fig. 4.15 Alternative dimensioning

is found and used as the second datum, which ensures symmetry (i.e. the machined boss faces will stand out equal amounts from the casting).

d) Again, the marking-out method given at (c) above is in accord with the drawing specification, although this is not immediately obvious. Although not stated, the drawing implies equality of the boss faces about the work centre. If this is not clear, then it must be shown on the drawing, as indicated in fig. 4.15.

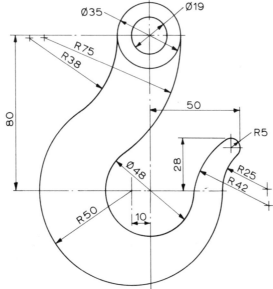

Fig. 4.16

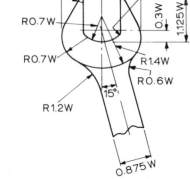

Fig. 4.17

Pitch line

90°

Axis of bolt

Design form of external thread
Fig. 4.18 (maximum-metal condition)

Exercises 4

1 Draw full-size the crane hook shown in fig. 4.16, showing clearly all constructions to obtain the centres of the circular arcs. Suggest and justify a suitable material and any precautions to be taken during manufacture. UEI

2 The proportions of one end of a spanner are shown in fig. 4.17. Draw full-size a spanner-end to these proportions when *W* is 40mm. Show all construction lines. Suggest a suitable material. UEI

3 Figure 4.18 shows the design profile of ISO metric external threads. Draw accurately, to an enlarged scale of 25:1, the given figure. The drawing should show a 0·3mm tolerance at the crest and root of the form, to ensure clearance in use. NCTEC

4 Draw full-size a view of the quadrant support bracket shown in fig. 4.19. Leave all construction lines in, so that the method of construction is clear. Also project a section taken through the bracket along line AA. NCTEC

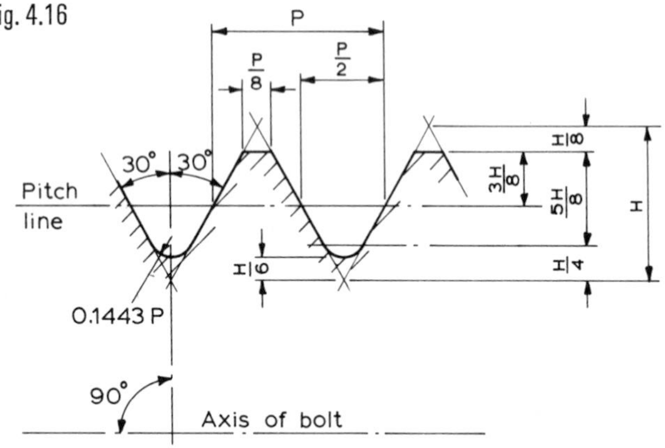

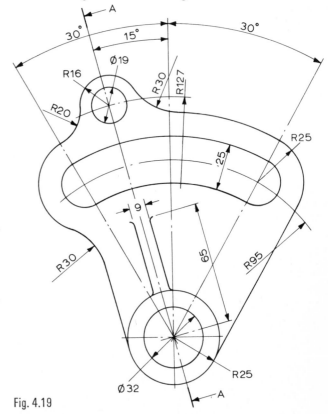

Fig. 4.19

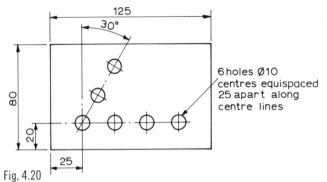

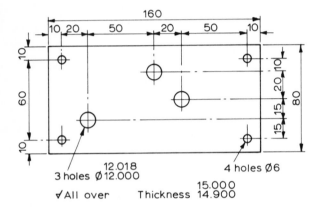

6 holes Ø10 centres equispaced 25 apart along centre lines

Fig. 4.20

Fig. 4.21 Mild-steel cover-plate

12.018
3 holes Ø 12.000

4 holes Ø6

15.000
Thickness 14.900

✓All over

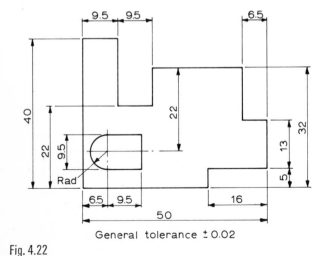

General tolerance ± 0.02

Fig. 4.22

5 Figure 4.20 shows a plate having a series of accurately spaced holes drilled through it. Redraw the component full-size. The drawing must be fully dimensioned with reference to hole A, using progressive coordinate dimensioning in accordance with BS 308.　　　NCTEC

6 A mild-steel cover-plate is shown in fig. 4.21.

a) By reference to the dimensioning method used for the holes, explain the term 'accumulation of tolerances', and show on a sketch a method of dimensioning which would avoid this problem.

b) With the aid of sketches, explain how a vernier height-gauge could be used for marking out the hole positions before drilling.

State why this method of locating the hole positions is suitable for a small batch of components, and suggest a method which may be more suitable for production of large quantities of the plate.　　　EMEU

7 Figure 4.22 shows a profile plate which is to be manufactured from 5 mm thick × 50 mm × 40 mm ground flat stock (gauge plate). With the aid of sketches, describe a method of marking out this plate. Assume all edges are square to each other.

8 Describe a suitable procedure for marking out the casting illustrated in fig. 4.23 prior to machining.　　　ULCI

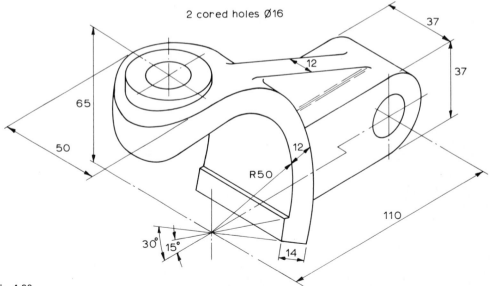

2 cored holes Ø16

Fig. 4.23

53

Chapter 5
Assembly of parts

1 Hand tools and processes

It is not the intention here to catalogue descriptions of hand processes and tools and the skills required to use them; in any event, it is no more possible to learn the skills of filing, say, from a book than it is to learn to swim. Therefore, we will confine ourselves to stating the advantages and limitations of the more common hand processes, and give a brief examination of each of these processes.

ADVANTAGES OF HAND PROCESSES

a) Very accurate when carried out by a highly skilled craftsman, although in general not so accurate as machining operations.
b) Extremely versatile, portable, and flexible, in that hand processing can be taken to and carried out at any site on a wide range of different jobs – hence indispensable in plant-maintenance work.
c) No power supply required except that of the human operator.
d) Tools and cutters relatively cheap.
e) Very suitable for a one-off assembly job where a degree of fitting is necessary to achieve the required fits.

LIMITATIONS OF HAND PROCESSES

a) Slow and therefore generally not suitable for quantity production work.
b) Inconsistent results must occur from time to time, due to the human influence.
c) A size restriction is present on the work, as the process and tool must be within the human size-range.
d) Hand processes (with the exception of the abrading processes such as lapping or polishing) are not suitable for relatively hard work materials.

The hand processes considered below are among the more important in use.

FILING This is the most widely used and versatile hand process. A large range of shapes and finishes can be produced when filing is used in conjunction with the marking-out process, and it is indispensable for chamfering and de-burring. It is possible by filing to match the degree of accuracy which can be achieved by any of the machining processes detailed in Chapter 1.

It is dangerous to use a file on work which is rotating in a centre lathe.

SCRAPING Scraping is a finishing process used mainly for achieving true flatness or roundness. The flat surface of a surface plate or machine slideway (see section 3.5) should ideally be scraped, since a 'non-wringing' surface (as opposed to a wringing surface – see section 3.6) is produced, this being advantageous. Flat scraping may be carried out by 'pull' or 'push' methods, some of the most accurate machine slides being processed by pull scraping. Round scraping is mainly confined to journal bearings into which shafts are fitted or 'bedded'.

CHISELLING Metalwork or 'cold' chisels are used in conjunction with a hammer, as are similar tools such as punches or drifts. Chiselling removes metal at a much faster rate than does filing or scraping, and is used for cutting flat or curved surfaces, slots, grooves, keyways, holes, etc. It is used mainly as a roughing operation.

SAWING A hacksaw is the main tool used for cutting off, though 'hack' is an unfortunate word to apply to a hand saw which is capable of achieving a similar precision to that of a machine saw. A hacksaw is also used for thin cuts such as are required for screwdriver slots or undercuts.

Common secondary hand processes applied to internal or external circular surfaces are reaming and threading by means of taps or dies. With care, the results are comparable to those produced upon a machine tool.

5.2 Combined hand and machine processes

Much work carried out in tool-rooms, maintenance departments, etc. is of the type which lends itself to a combination of hand and machine processes. In this class of work, the numbers required are usually small (often only one off), and accuracy is usually more important than speed. Hand processes then, have an important part to play in this field, which we will call *unit-batch manufacture*.

A second class of work where hand processing can be important is in assembling machined parts into a finished or part-finished unit. This can be seen on production lines where batches of moderate size are being handled, the operators and equipment being sufficiently flexible to cope with changes in the type of unit being manufactured. We will call this class of work *small-batch assembly*.

Let us consider an example from each of the classes of work described above. In both examples, a sequence of operations is given for a specified component.

Unit-batch manufacture

Assume one component is to be produced as drawn in fig. 5.1. Only standard machines, cutting tools, hand tools, and marking-out tools are available, as described in Chapters 1, 2, 3, and 4. A planning sheet is shown in fig. 5.2, which lists one possible sequence of operations for the manufacture of this part. Although no geometrical tolerances are specified for flatness, straightness, or parallelism (these factors not being of crucial importance), it will nevertheless assist subsequent operations if these geometrical factors are carefully controlled in the initial preparation of the part.

Small-batch assembly

In this example we will assume that occasional batches of 100's of the sub-assembly shown at fig. 5.3 are required. Further, we will not detail the machining methods used to manufacture the individual parts, but will merely list the sequence of operations necessary to assemble the separate component parts into the finished unit. A planning sheet is shown at fig. 5.4, giving a possible assembly sequence using hand methods.

In much hand-assembly work, aids such as special hand tools, holding or locating devices, hand presses with special tools, etc. are used to facilitate assembly.

5.3 Assembly drawings

Drawings may be broadly grouped into two categories, viz. *assembly drawings* and *single-part drawings*. Examples of both have been shown in earlier sections of this book, but we will consider assembly drawings at greater length here, and single-part drawings in section 5.4. First, it might be useful to state the recommended drawing-sheet sizes. The British Standards Institution has recommended that the International Standards Organisation (iso) 'A' series be adopted. The common range of sizes to be used from this series is given below:

Designation	Sheet size (mm)
A0	841 × 1189
A1	594 × 841
A2	420 × 594
A3	297 × 420
A4	210 × 297

Fig. 5.1 Component drawing

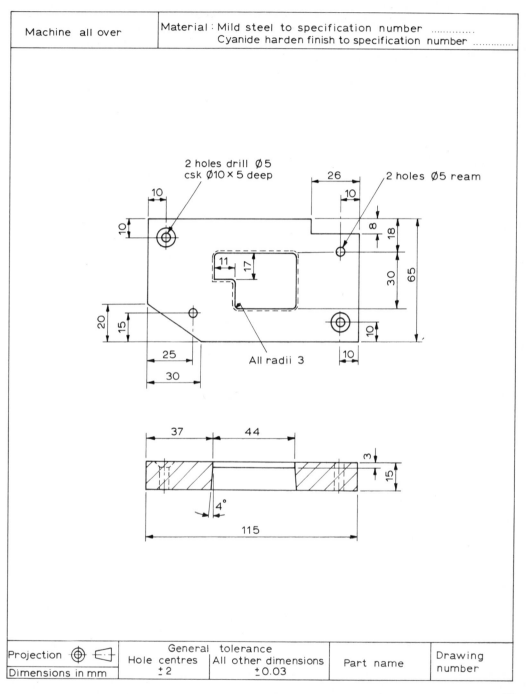

Machine all over

Material: Mild steel to specification number ....
Cyanide harden finish to specification number ....

2 holes drill Ø5
csk Ø10 × 5 deep

2 holes Ø5 ream

All radii 3

Projection ⊕ ⊏⊐		General tolerance		Part name	Drawing number
Dimensions in mm	Hole centres ±2	All other dimensions ±0.03			

Opn no.	Sketch of operation	Operation	Machine used	Tool used
1		Saw 20×70 black mild steel to 120 long.	Hacksawing machine	
2	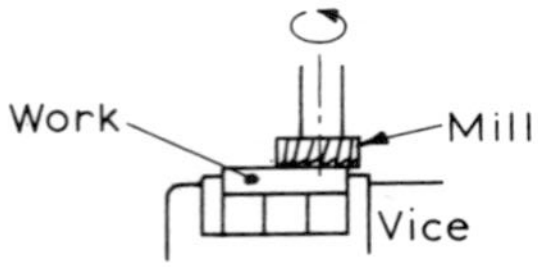	Mill opposite faces flat and parallel to 15 thick.	Vertical milling machine	Face mill
3	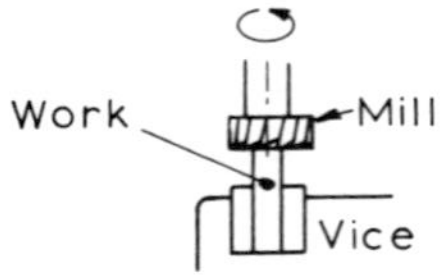	Mill four edges flat, square and parallel to 115×65.	Vertical milling machine	Face mill
4		De-burr		Hand file
5	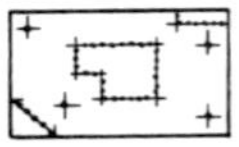	Mark out on front face. Dot-punch profiles and hole centres.		Surface plate, Angle plate, Vernier height-gauge, Scriber, Rule, Dividers, Dot punch
6	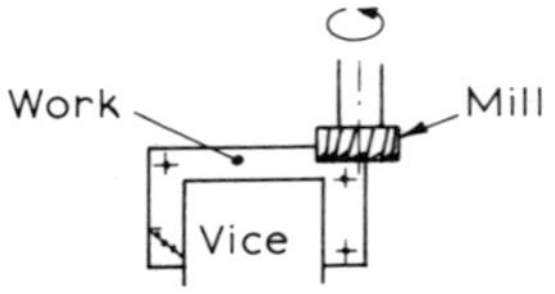	Mill step and angular face.	Vertical milling machine	Face mill
7	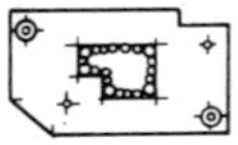	Drill holes for reaming. Drill and csk 2 holes Ø5. Drill Ø6 holes at profile corners. Drill Ø2 holes around profile.	Sensitive drilling machine	Twist drills
8		Rough cut central profile to shape.		Flat chisel, Hammer
9		Finish file central profile to shape. Back off to 4°.		Flat file
10		Ream Ø5 holes.		Hand reamer, Wrench
11		De-burr.		Hand file, Half-round scraper
12		Final inspection.		Rule, Try square, Vernier caliper, Caliper micrometer, Protractor, Radius gauge

Fig. 5.2 Planning sheet

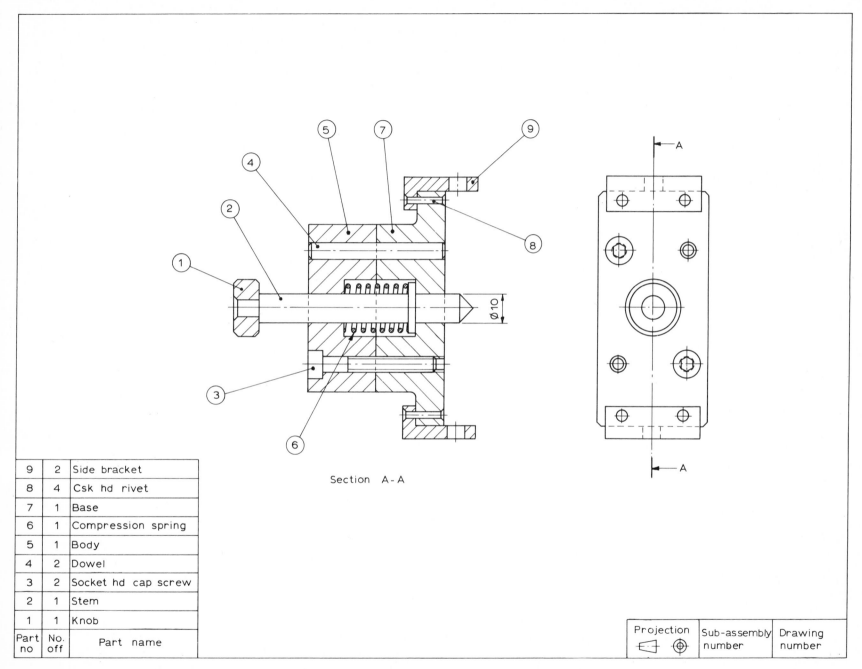

Part no	No. off	Part name
9	2	Side bracket
8	4	Csk hd rivet
7	1	Base
6	1	Compression spring
5	1	Body
4	2	Dowel
3	2	Socket hd cap screw
2	1	Stem
1	1	Knob

Section A-A

Projection	Sub-assembly number	Drawing number

Fig. 5.3 Assembly drawing

OP'N NO.	OPERATION	TOOL USED
①	De-burr all parts.	Hand file Half-round scraper
②	Provisionally locate parts ⑤ & ⑦ together with dowels part ④. Run ∅10mm reamer thro' bore.	∅10mm hand reamer Wrench
③	Dis-assemble above. Thread ⑥ over ② and locate in ⑦.	
④	Assemble ⑤ to ⑦ with ④. Screw up with ③.	Light mallet Allen key
⑤	Assemble ① to ② and rivet.	Base-plate with ∅12mm hole Punch Hammer
⑥	Assemble ⑨ to ⑦, placing ⑧ in position. Rivet.	Plate Punch Rivet
⑦	Lightly grease.	
⑧	Inspect.	

Fig. 5.4 Planning sheet

All drawing sheets should have a *title block*, preferably located at the bottom right-hand corner, which generally should contain at least the following information:
a) name of company,
b) drawing number,
c) drawing title,
d) scale,
e) date,
f) name of draughtsman,
g) 'issue' information,
h) projection symbol (first- or third-angle)
i) unit of measurement for dimensions.

An *assembly drawing* shows the various component parts arranged together or assembled as a complete finished unit. Overall dimensions are sometimes included if convenient, but the detailed dimensioning of the individual parts is reserved for the single-part drawings.

Sub-assembly drawings may be produced from the assembly drawing, showing how individual parts fit together into smaller units which in turn form part of the complete general arrangement. Sectional views are greatly used in all assembly drawings, to facilitate clear depiction of mating parts. Individual parts are identified on assembly drawings by means of a '*balloon-reference*' system in which lines are drawn from each part to a circle which encloses the part number (see figs 5.3 and 5.5). A *parts list* is included with the title block on assembly drawings, as shown in fig. 5.5, but in larger designs the parts may be listed upon a separate sheet. A parts list should contain the following information:
a) part number,
b) part description,
c) quantity required (no. off),
d) part-drawing number.

Note with respect to (d) above that, in the case of a standard part, the number of the standard drawing will be referred to as a cross-reference; hence it is not necessary continually to duplicate similar part drawings for different designs.

Figure 5.5 shows an assembly drawing of a hand-operated clamp; the balloon references and parts list can be seen in the figure. Also, a list for revisions has been included at the bottom left-hand corner. It is essential with all drawings that any modifications or revisions are always added to the original drawing. If a revision were made in the example shown in fig. 5.5, then the drawing number would change to FG 55/A, and so on. This ensures that every drawing in an engineering works depicts exactly the part(s) it represents as it actually is. In the author's experience, this important feature of drawing-office routine is often sadly neglected, leading to much unnecessary error and expense.

Figure 5.5 also illustrates the fact that it is not always necessary to 'cross-hatch', or line in, every part that is sectioned. Standard parts such as dowels, screws, spindles, etc. are left plain for clarity, i.e. they are not shown in section. This is also conventional practice for ribs of castings, as illustrated.

5.4 Single-part drawings

As with assembly drawings, all *single-part drawing* sheets should bear a *title block*. Single-part drawings present a complete specification of each individual part, including fully dimensioned views with appropriate machining instructions. The assembly drawings (including parts list), sub-assembly drawings (where required), single-part drawings, and planning sheet (see section 2.5) constitute a complete *job specification*, giving all the necessary information for production. It will often be found convenient to use a separate detail sheet for each individual part, so that separate prints can easily be produced and distributed to various departments if necessary. The A4 sheet is a useful size for this purpose.

In addition to the title block, a single-part drawing should also list the following information:

a) material specification,
b) heat-treatment specification,
c) finish required,
d) general tolerance note, where required,
e) any other general notes regarding machining, fits, tool references, etc., where thought necessary for clarity.

Figures 5.6(a) to (c) show single-part drawings of the non-standard parts for the clamp assembly drawn in fig. 5.5.

Everything contained in these single-part drawings has been explained earlier in this book, with the exception of the method used to designate *threads*. Figure 5.3 shows how assembled screws are depicted, and fig. 5.5 shows how threads are drawn in section. When ISO metric screw-threads are required, they should be specified on a drawing in the manner shown in fig. 5.6(a). This thread form is covered in BS 3643: parts 1, 2, and 3, 'ISO metric screw-threads', which recommends that the *coarse*-pitch series is used for the vast majority of general-purpose applications, and the *fine*-pitch

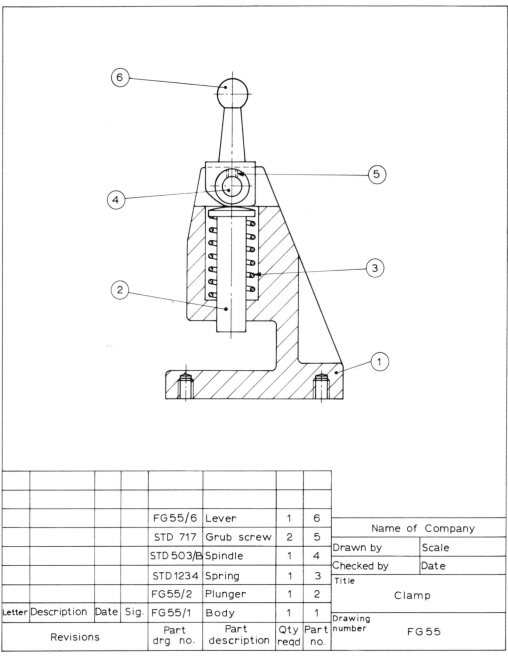

				Part drg no.	Part description	Qty reqd	Part no.		
				FG55/6	Lever	1	6	Name of Company	
				STD 717	Grub screw	2	5		
				STD 503/B	Spindle	1	4	Drawn by	Scale
				STD 1234	Spring	1	3	Checked by	Date
				FG55/2	Plunger	1	2	Title	
Letter	Description	Date	Sig.	FG55/1	Body	1	1	Clamp	
Revisions				Part drg no.	Part description	Qty reqd	Part no.	Drawing number	FG55

Fig. 5.5 Assembly drawing

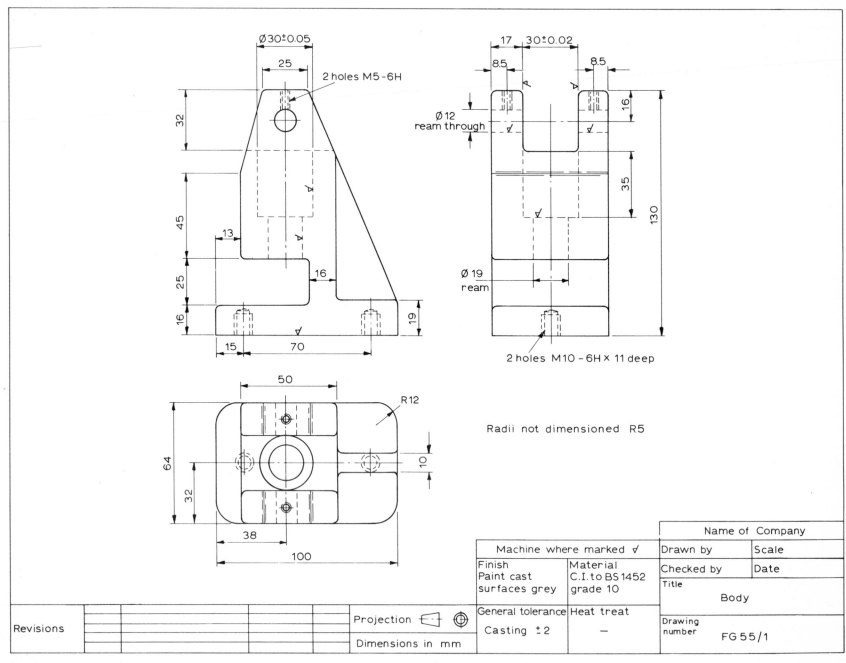

Ø30±0.05
25
2 holes M5-6H
32
45
13
25
16
16
15
70
19

17   30±0.02
8.5   8.5
Ø12
ream through
16
35
130
Ø19
ream
2 holes M10-6H × 11 deep

50
R12
64
32
10
38
100
Radii not dimensioned R5

Revisions							Projection ⊏⊐ ⊕	Machine where marked √		Name of Company	
									Drawn by	Scale
							Dimensions in mm	Finish Paint cast surfaces grey	Material C.I. to BS 1452 grade 10	Checked by	Date
										Title Body
								General tolerance Casting ±2	Heat treat —	Drawing number FG 55/1	

Fig. 5.6(a)   Single-part drawing

60

series for all other applications. The standard shows three alternative classes of fit for coarse and fine threads, which are designated thus:

Class of fit	Tolerance class		Recommended use
	Internal	External	
Close	5H	4h	High-quality production
Medium	6H	6g	Most general engineering purposes
Free	7H	8g	Quick and easy assembly

All ISO threads are designated on drawings by the letter M, followed by the diameter (mm), the pitch (mm) in the case of fine threads only, and finally the thread-tolerance symbol, according to the above table.
The following two examples illustrate the method:
a) M5–6H indicates a coarse-series internal thread of 5mm diameter, being a medium-class fit;
b) M12×1·25–6g indicates a fine-series external thread of 12mm diameter × 1·25mm pitch, being a medium-class fit.

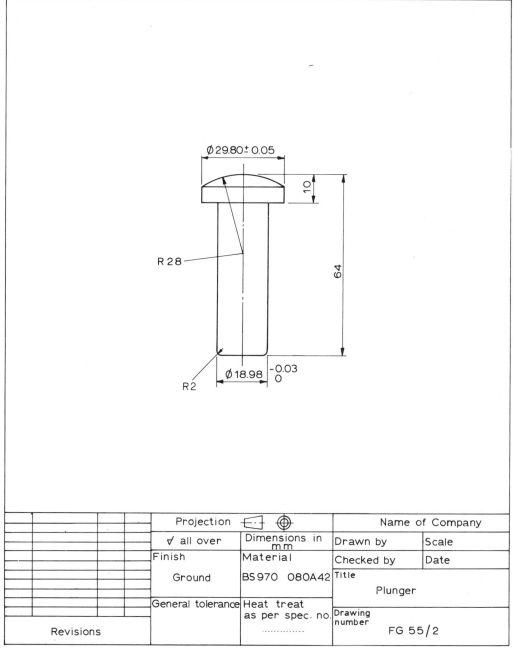

Fig. 5.6(b) Single-part drawing

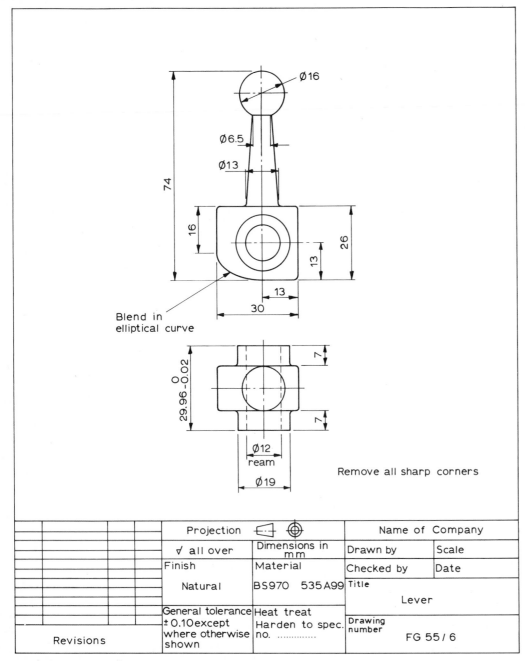

Fig. 5.6(c) Single-part drawing

5.5 Tolerance accumulation

Tolerance accumulation, or tolerance build-up, as it is sometimes called, refers to the accumulation of individual tolerances on lengths, which in turn affects the allowable tolerance which can be placed on the overall length. The effect of a cumulative tolerance must be taken into account both on overall lengths of individual components and on the overall length of assembled parts.

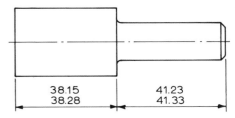

Fig. 5.7 Part dimensioned for length

Consider the part shown dimensioned in fig. 5.7. The build-up of tolerances over the whole length is equal to the sum of the individual tolerances, and is $0\cdot13+0\cdot10 = 0\cdot23$ mm; therefore the overall length cannot be allowed to vary by more than this amount. Its largest value will be $41\cdot33+38\cdot28 = 79\cdot61$ mm, and its smallest value will be $41\cdot23+38\cdot15 = 79\cdot38$ mm, the difference being $0\cdot23$ mm.

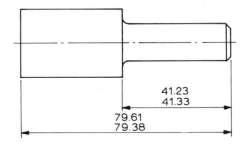

Fig. 5.8 Part dimensioned for length

For the purposes of manufacture, it might be more convenient to dimension the part in the alternative manner shown in fig. 5.8. In effect, the right-hand face of the part as drawn has become the datum face. With the overall length toleranced as shown, and allowing for the cumulative effect, it ensures that the 38·15/38·28 mm dimension will be maintained. This is not so if any other tolerances than those shown are placed upon the overall length.

A similar effect occurs in the assembly of individual parts where the overall length is the sum of the individual part lengths. Consider the assembly shown in fig. 5.9; it can be seen that

overall basic length = sum of individual part basic lengths

i.e.
$$L = X + Y + Z$$

and cumulative tolerance = sum of individual tolerances

i.e.
$$+q = (a+b) + (c+d) + (e+f)$$

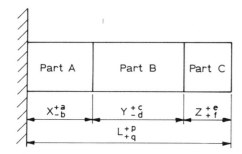

Fig. 5.9 Assembly of parts

As stated earlier, this is true not only for assemblies but also for accumulated tolerances on individual components. It can also be stated that calculations of cumulative tolerances are much simplified if a common method of specifying limits is used throughout a design; for example, the method of tolerancing shown in fig. 5.10(a) is preferable to that shown in fig. 5.10(b).

It can be seen from fig. 5.10(a) that

overall basic length = 50+72+36 = 158 mm
and cumulative tolerance = (2+1)+(5+3)+(1+3) = 15 mm

A note of caution must be sounded here, following the somewhat simplified introduction to the topic of tolerance accumulation. In practice, a sample of parts chosen at random to form an assembly, for example, will only very infrequently all be at the maximum length (or minimum length) condition, hence giving the largest (or smallest) assembly. In other words, the extreme condition happens rarely, and the average condition happens most often. This, therefore, is a statistical matter requiring additional study not appropriate at this stage.

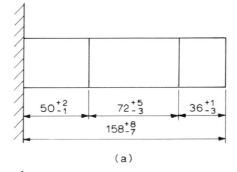

(a)

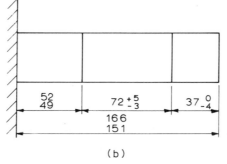

(b)

Fig. 5.10 Toleranced dimensions

Exercises 5

1 Figure 5.11 gives the details of two rod ends which are to be conjoined by means of the cotter and gib shown. Make a longitudinal sectional drawing of the assembly, scale full-size. The section is to be on the plane AA. Itemise each detail on the drawing, and add a title block with a parts schedule.　　　　　NCTEC

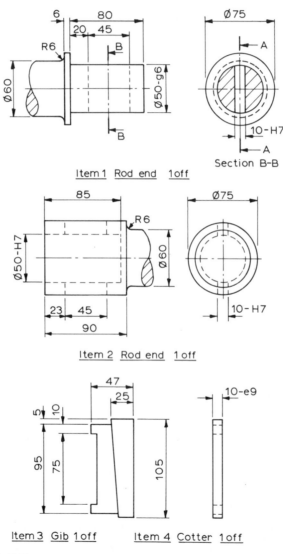

Item 1　Rod end　1off

Section B-B

Item 2　Rod end　1off

Item 3　Gib　1off　　　Item 4　Cotter　1off

Fig. 5.11

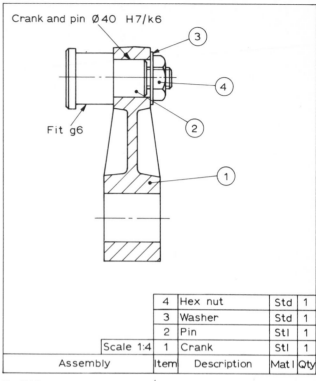

Crank and pin Ø40　H7/k6

Fit g6

4	Hex nut	Std	1
3	Washer	Std	1
2	Pin	Stl	1
Scale 1:4　1	Crank	Stl	1
Assembly	Item	Description	Matl　Qty

Fig. 5.12

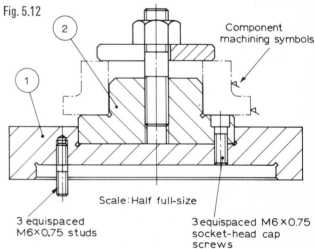

Component machining symbols

Scale: Half full-size

3 equispaced M6×0.75 studs

3 equispaced M6×0.75 socket-head cap screws

For detailing purposes dimensions should be scaled. Tolerances to BS 4500.

Fig. 5.13

2 Figure 5.12 shows a sectional assembly drawing through a crank lever and pin, scale quarter full-size. On a standard drawing sheet, prepare a fully dimensioned single-part drawing of the pin, scale full-size.   NCTEC

3 The sectional drawing shown in fig. 5.13 is an assembly drawing of a turning fixture. On a standard drawing sheet, prepare fully dimensioned single-part drawings of (a) the base plate, item 1, (b) the location spigot, item 2.
NCTEC

4 Figure 5.14 shows a designer's sketch of a counterboring cutter, part-sectioned to show how the cutter and pilot are driven and retained in the body. On a standard drawing sheet, draw items 1, 3, and 4 separately, using the minimum of views consistent with the given information. Complete the parts list and title block.

Any feature of design or dimensions that is not clearly shown should be sensibly interpreted.   NCTEC

5 Figure 5.15 shows an assembly of a simple flange-coupling and adaptor. By scaling the drawing, make a fully dimensioned drawing of the parts X and Y. Insert limits of size in accordance with the BS 4500 primary selection of fits, where necessary.

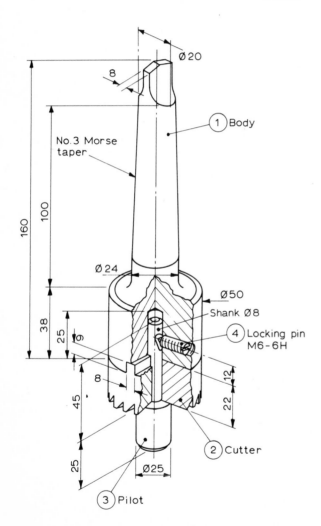

Fig. 5.14

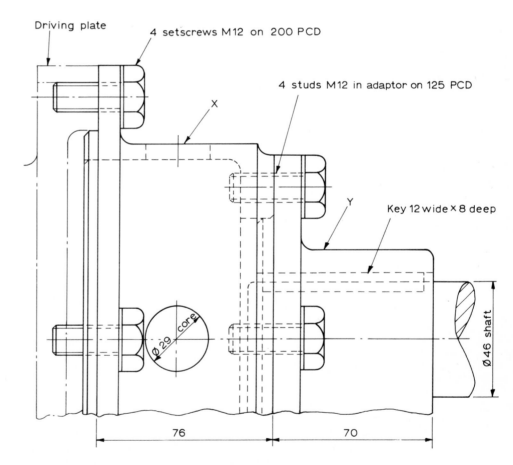

Fig. 5.15

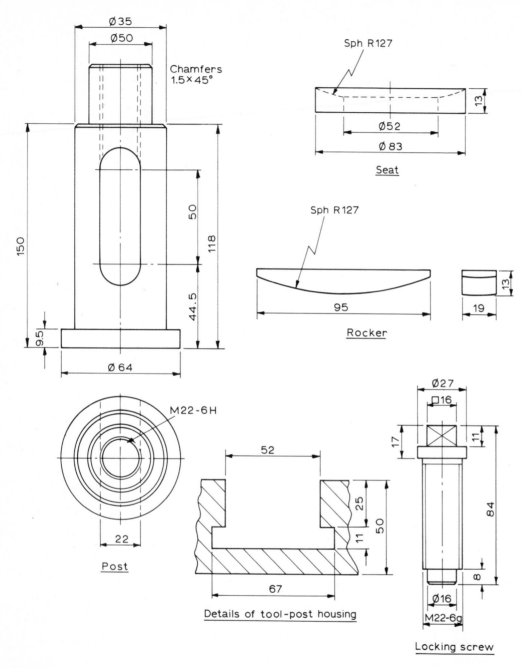

Ø35

Ø50

Chamfers
1.5×45°

Sph R127

Ø52

Ø83

Seat

Sph R127

150

50

118

44.5

9.5

Ø64

Rocker

95

13

19

M22-6H

Post

22

52

25

50

11

Details of tool-post housing

67

Ø27

□16

17

11

84

8

Ø16

M22-6g

Locking screw

6 Figure 5.16 shows details of a lathe tool-post and a section through the tool-post housing. On a standard drawing sheet, draw a full-size sectional elevation of all the parts assembled. Add a number reference to each part, and complete the parts list and title block in accordance with BS 308.   NCTEC

7 Figure 5.17 shows the details of a clapper-box and tool-post assembly for a shaping machine.

a) Draw, full-size, with all parts assembled, a sectional elevation on the plane XX. The edge of the cutting tool, item 8, is to be shown at a maximum distance of 50mm from the bottom of the swivel-plate.

b) Draw a front elevation in first-angle projection.

c) Add the following dimensions:
  i) horizontal overall length of sectional assembly,
  ii) maximum length of swivel-plate,
  iii) maximum distance of tool cutting edge from the bottom of the swivel-plate,
  iv) centre of tool-post to tool point,
  v) the tool-post diameter.

Show, and give suitable values to the angles known as top rake, front cutting-edge clearance, and approach angles. These should be suitable for the material to be cut.

d) Make a parts list with information, including materials, to enable parts to be ordered and assembled. Use 'balloons' to identify the components on the drawing.

e) Add a title and scale, and state the projection used.   EMEU

Fig. 5.16

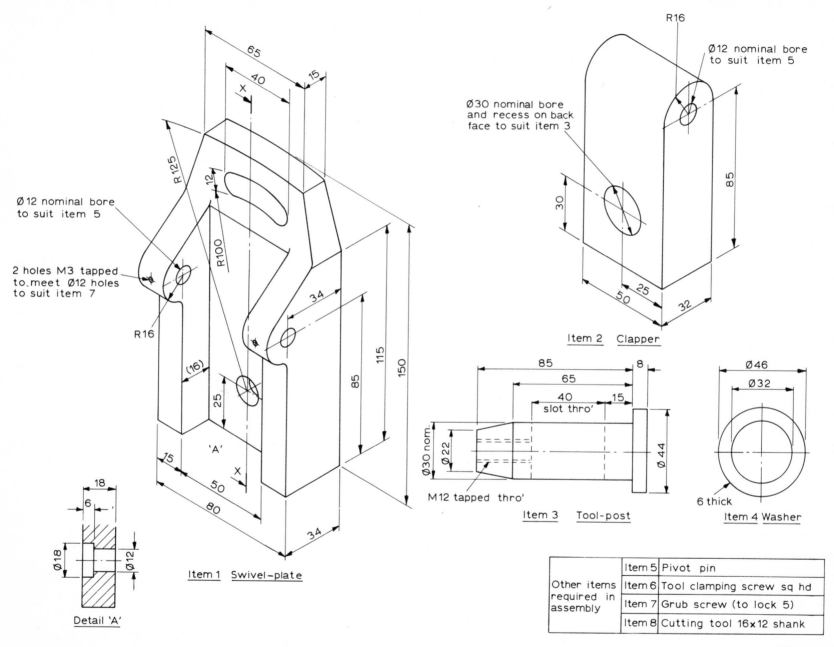

Ø12 nominal bore
to suit item 5

2 holes M3 tapped
to meet Ø12 holes
to suit item 7

R16

(16)

25

'A'

15

50

80

34

18

6

Ø18

Ø12

Detail 'A'

65

40

X

15

R125

12

R100

34

115

150

85

X

Item 1   Swivel-plate

R16

Ø12 nominal bore
to suit item 5

Ø30 nominal bore
and recess on back
face to suit item 3

30

85

50

25

32

Item 2   Clapper

85

65

40

15
slot thro'

8

Ø30 nom.

Ø22

Ø44

M12 tapped thro'

Item 3   Tool-post

Ø46

Ø32

6 thick

Item 4 Washer

Other items required in assembly	Item 5	Pivot pin
	Item 6	Tool clamping screw sq hd
	Item 7	Grub screw (to lock 5)
	Item 8	Cutting tool 16×12 shank

Fig. 5.17

67

8 Figure 5.18 shows the two main items for a milling fixture. The component also shown is supplied as mild-steel bright bar, and is to have the flats machined by straddle milling, after correct location and clamping in the fixture.

a) Draw, full-size, the following views of the fully assembled fixture:

 i) an elevation viewed as arrow A,

 ii) a sectional plan with the 'cutting plane' through the clamping studs.

In each view, the component is to be shown correctly secured; clamping studs, nuts, and washers of suitable proportions must therefore be added. Also show standard holding-down bolts, and include part of the machine table, to indicate their location.

All other information not supplied is to be decided. No dimensions need be added, and hidden detail is not required.

b) Make a parts list of the details, including materials. Identify the details on the assembly by the use of 'balloons'.

c) Add a title and scale, and state the method of projection used.
EMEU.

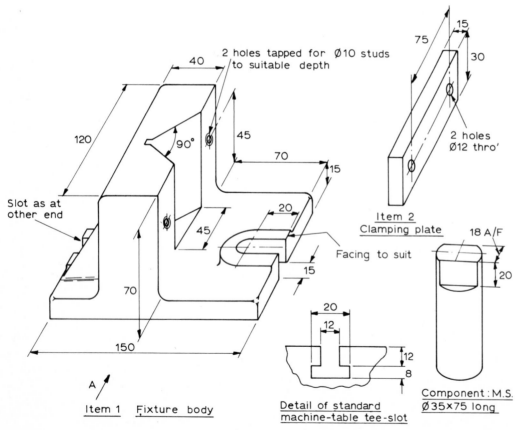

Item 1   Fixture body

Item 2
Clamping plate

Detail of standard
machine-table tee-slot

Component : M.S.
Ø35×75 long

Fig. 5.18

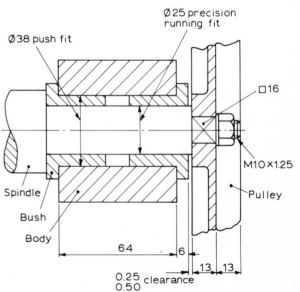

Fig. 5.19

9 Figure 5.19 shows a sectional view of a pulley assembly. Draw the single-part drawings of the following: (a) the body, (b) the flanged bushes, (c) the spindle.   UEI

10 The details of a spring-stop are shown in fig. 5.20. On a standard drawing sheet, draw, full-size, a sectional elevation showing all the parts assembled.

Add a number reference to each part, and complete the parts list and title block in accordance with BS 308.
     NCTEC

11 Assume one lathe tool-post is to be produced as drawn in fig. 5.15, and only standard machines and tools are available, as described in Chapters 1, 2, 3, and 4. Draw up a sequence of operations for the production of each individual part shown.

12 Draw up a sequence of operations for the production of items 1, 2, 3, and 4 from fig. 5.17, given the same conditions and limitations described in question 11. Also draw up a sequence of operations for the assembly of the complete clapper-box and tool-post.

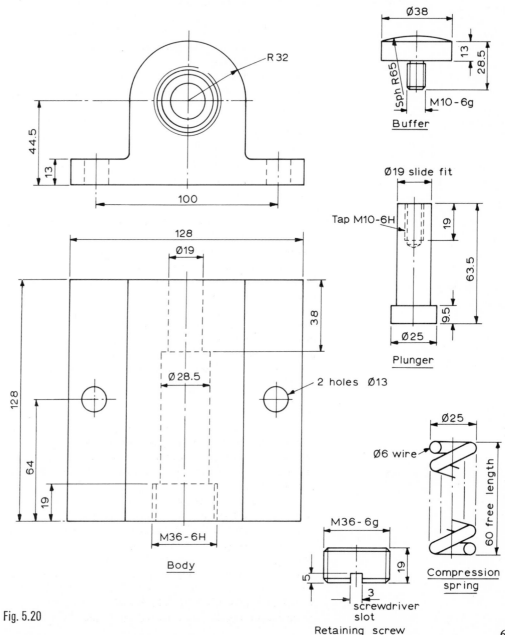

Buffer

Plunger

Body

Retaining screw

Compression spring

Fig. 5.20

69

Chapter 6 Materials

1 Properties of materials

The *physical properties* of materials used for engineering purposes are those properties of conductivity of heat and electricity, melting-temperature, density, magnetism, etc. However, it is the mechanical properties of engineering materials with which we are concerned here, because it is particularly upon these properties that the usefulness or otherwise of materials depends.

The major mechanical properties which are of interest to engineers are *hardness*, *toughness*, *ductility*, and *malleability*.

HARDNESS is a measure of the resistance of a material to scratching, abrasion, or indentation by a harder body; i.e. the harder a material, the more resistant it is to wear. Hardness is linked to the physical properties of metals in that generally the harder a metal, the higher its temperature must be raised in order to melt it.

TOUGHNESS is a measure of the amount of energy a material can absorb during bending or shearing without fracture, i.e. a very tough material can withstand a large bending force before it breaks–note that a very hard and brittle material might not, because a hard material is not necessarily also tough.

DUCTILITY refers to the capacity of a material to be drawn out in tension without rupture. This property in a material is essential in such engineering processes as wire-drawing, tube-extrusion, and cupping and deep-drawing of sheet metal.

MALLEABILITY is similar to ductility, but refers to the capacity of a material to be squeezed out in compression without rupture. This property is associated with such engineering processes as drop-forging, hammering, and rolling. A ductile material is usually malleable, but this is not always so.

One of the difficulties we must face here is that none of these basic mechanical properties can be expressed in terms of a simple numerical value. Furthermore, an engineer is usually most concerned with the magnitude of the forces which cause deformation, not the type of deformation itself. The designer, when 'calling for' a material on a drawing, will wish precisely to specify a material in such a way that he will be confident that it will possess the correct mechanical properties to withstand the various forces to be imposed upon it during service. Therefore, it has become necessary to evolve certain mechanical tests related to mechanical properties, the results of which can be expressed in numerical terms. Many tests have now become standard practice, the results being widely regarded as criteria (standards of judgement) for choosing a material. If one consults BS 970, 'Wrought steel' it will be found that the mechanical properties are specified in the form of tensile strength, yield stress, elongation %, and Izod and Brinell values. These numerical values, which directly or indirectly relate to the basic mechanical properties, are all derived from mechanical tests.

It is not appropriate to study the equipment, principles, and results of such tests here, but we will briefly refer to the more important tests, viz. the hardness test, the tensile test, and the impact test, in order to appreciate the numerical units used.

Hardness test

There are several hardness tests in use, each having its own particular suitability for different classes of work, for example the Brinell test (see BS 240), the Vickers pyramid hardness test (see BS 427), the Rockwell test (see BS 891), and the Shore scleroscope test. Each of these tests results in a number which enables the relative hardness of different materials to be compared. Each of the test numbers is loosely related to another (see BS 860), and hardness numbers may be roughly related to mechanical properties other than hardness, such as ductility for example.

The *Brinell test* is the best known of the harness tests, being described in BS 240. In the test, a hardened steel ball is forced into the surface of the workpiece by the application of a suitable standard load. When the diameter of the impression has been measured, the Brinell hardness number (BHN) can be calculated thus:

$$\text{BHN} = \frac{\text{applied load (kg)}}{\text{surface area of impression (mm}^2)}$$

Tensile test

In this test, a standard test-piece of material is subjected to a tensile force, and will therefore reveal information about the strength of the material in tension and its ductility. (Conversely, a compressive force could be applied, hence yielding information about malleability.) The recommended method of carrying out this test is outlined in BS 18. Under load, the circular test-piece elongates before breaking, the increase in length, reduction in diameter, and forces applied being carefully recorded. The test results can be plotted, to give the well known force–extension diagram, and also used to give the following data.

a) Tensile strength (N/mm²) $= \dfrac{\text{maximum force (N)}}{\text{original cross-sectional area (mm}^2)}$

b) Yield stress (N/mm²) $= \dfrac{\text{yield force (N)}}{\text{original cross-sectional area (mm}^2)}$

c) Percentage elongation (%) $= \dfrac{\text{extension of gauge length (mm)} \times 100}{\text{original gauge length (mm)}}$

d) Percentage reduction in area (%) $= \dfrac{\text{maximum decrease in area (mm}^2) \times 100}{\text{original cross-sectional area (mm}^2)}$

Impact test

This test enables the shock-resistance of a material to be assessed by subjecting a notched specimen to the impact of a single blow. The result is therefore related to the toughness of the specimen, but must be interpreted with care. The test is usually carried out upon an Izod or Hounsfield impact-testing machine.

An Izod value is a measure of the energy which a material test-piece can absorb, and will be expressed in the energy units of joules (J). An impact test is very useful for revealing the degree of brittleness in a material due to faulty heat-treatment, this state not being revealed by either the hardness or the tensile test. Note that the Izod test is for the most part being superceded by the Charpy test for material specification purposes (see BS 131).

Machinability

Before we leave this section concerned with the properties of materials, we should consider one other factor which may be of importance in the specification of metals. This factor is called the 'machinability' of a metal, and is of particular interest to a production engineer, as it refers to the ease with which a workpiece may be processed upon a machine tool. Of course, the machining techniques used, such as the type of machine, speeds, feeds, cutting-tool shape (see section 2.1), etc. will all affect the ease with which metal can be sheared from a workpiece; however, the structure of the material can also influence machinability, and it is this metallurgical aspect with which we are concerned here. In general, a very ductile material does not

machine as well as a more brittle material, since the chips do not so easily fracture and shear. There are three ways in which the structure of a material may be modified in order to make it more amenable to machining, as follows.

a) BY INTRODUCING A FREE-CUTTING ELEMENT Lead is commonly added to ferrous metals (such as free-cutting mild steel, 'ledloy' steel, or stainless steel) or non-ferrous metals (such as brass or bronze). It exists as particles of the pure element randomly distributed throughout the structure. These particles cause discontinuity in the structure, thus assisting the chips to shear more freely from the work. Friction is reduced, which leads to less power being needed and less tool wear. From 1% to 3% lead is added to copper-based alloys (see BS 249), but only up to 0·35% is added to steel.

Sulphur is also used in steel for the same purpose, and combines with manganese in the steel to form manganese sulphide. This exists in the structure as isolated globules, which have a similar effect to that of lead particles. Up to 0·25% sulphur is added to steel, in conjunction with 1% to 1·5% manganese. The full specification range of free-cutting steels can be found in BS 970.

b) BY HEAT-TREATMENT Steels are heat-treated in order to ensure that the structure is in the most receptive condition for machining.

Low-carbon plain steel has better machinability in the normalised condition. Normalising is carried out by heating the steel to a temperature approximately 30°C above the upper critical temperature (which varies with the carbon-content) and then cooling in air.

High-carbon plain steel has better machinability in the annealed condition. Annealing is carried out by heating the steel to a temperature approximately 30°C above the lower critical temperature (695°C) and allowing it to cool slowly in the heat-treatment furnace.

c) BY COLD-WORKING A cold-worked material is less ductile than a hot-worked one, and should therefore machine more freely. It is said that a cold-drawn ('bright') rod of metal has better machinability than a hot-rolled ('black') rod of similar metal, although the difference is probably marginal.

TESTING MACHINABILITY The concept of machinability

is not an easy one to describe or express in numbers. It has been suggested that the Brinell test gives a guide to the machinability of a metal, depending upon the cutting-tool material being used. For example, using high-speed steel tools, machining may become difficult when the BHN for the work material is above 300. Conversely, with a BHN below 100, the material may be too ductile to machine satisfactorily, and will tear rather than shear.

6.2 Ferrous metals

Metals which have iron (in latin, *ferrum*) as their main constituent are called *ferrous metals*, and include all the many different types of irons and steels. All ferrous metals originate from iron ore, from which iron is extracted in a blast-furnace. Figure 6.1 shows the relationship of the iron- and steel-making processes leading from the iron-ore stage to the finished products.

The iron from the blast-furnace is run off into moulds,

the bars formed from the moulds being called 'pigs'. *Pig iron* is an impure material, the properties of which are too poor to allow it to be used as an engineering material, but, as can be seen from fig. 6.1, it is the basic raw material from which all commercial irons and steels are produced. There is not space in a book of this kind to examine all the processes and materials referred to in fig. 6.1, and we shall consider only the compositions and properties of cast irons and plain-carbon steels.

Cast iron

Pig iron is refined in a foundry cupola, the product of which is *cast iron*. The molten metal from the cupola is tapped into ladles, or directly into moulds. The great majority of cast iron used in the engineering industry is of the type called grey cast iron (see BS 1452, 'Grey iron castings'). This is an alloy (mixture) of ferrite and carbon, with a carbon-content of about 3·5%. Relatively speaking, this is a large amount of carbon; so not all of it is able to combine with the iron, some of it existing in the structure in a free state. This free carbon acts as a lubricant, and also improves the machinability by causing discontinuity in the structure (having a similar effect to that caused by the introduction of lead into some metals, as described in section 6.1). The hardness of ordinary grey cast iron (about 250 BHN) allows it to be freely machined, but it can be made much harder by chilling, i.e. by inserting metal 'chills' in the mould to conduct the heat rapidly away from the molten iron during the cooling of the casting (see section 7.2).

A typical composition of ordinary grey cast iron is 94% ferrite, 3·2% carbon, plus small percentages of silicon, manganese, sulphur, and phosphorus.

This material is of the utmost importance to the engineering industry, and is used almost exclusively in the form of castings produced from sand moulds. This is because (a) it is cheap, (b) it has a low melting-temperature (1150–1200 °C), and (c) is very fluid in the molten condition, hence easily filling mould cavities.

Cast iron is about four times stronger in compression than in tension, and about twice as strong in bending as in tension. Obviously, therefore, a designer should always endeavour to eliminate tensile and bending forces from iron castings. Also, cast iron is brittle and easily broken under shock loads, but has the useful property of being able to absorb (or damp out) vibration more easily than other metals.

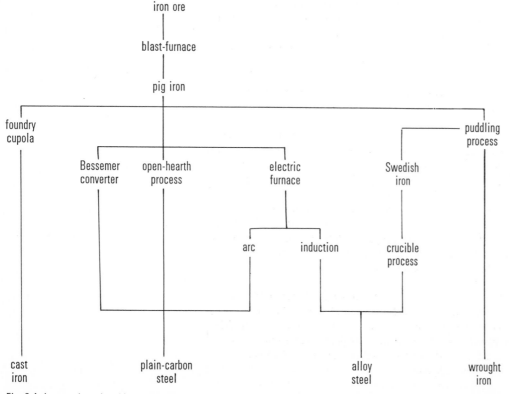

Fig. 6.1 Iron- and steel-making processes

To sum up, the general advantages and limitations of cast iron are as follows.

ADVANTAGES
a) Cheap
b) Easily cast into shapes–intricate or plain, large or small
c) Good machinability
d) Self-lubricating
e) Strong in compression
f) Vibration absorbent

LIMITATIONS
a) Brittle
b) Weak in tension
c) Suitable for processing only in the cast form
d) More difficult to weld than plain low-carbon steel

Many thousands of components are cast to shape in cast iron before being machined to their final form. Typical examples are machine-tool beds, engine cylinder-blocks, bearing brackets, brake drums, machine frames and housings, and pipes. Other types of cast irons are available (in addition to the plain cast irons) which have improved properties due to special heat-treatment or the addition of alloying elements. The most important of these are described below.

MALLEABLE CAST IRON (BS 309, 'Whiteheart malleable iron castings'; BS 310, 'Blackheart malleable iron castings'; and BS 3333, 'Pearlitic malleable iron castings') Malleable-iron castings, relatively speaking, are much more malleable than grey-cast-iron castings, and can withstand higher shock loads. The three types of malleable irons referred to above are all obtained by heat-treating brittle white-iron castings. All the carbon in white cast iron is combined with the iron to give iron carbide, no free carbon existing to darken the colour of the iron. The names of the *blackheart* and *whiteheart* processes refer to the colours of a fractured section after malleablising has been carried out. *Pearlitic* malleable iron structures show the carbon congregated in the form of 'rosettes', the castings possessing high tensile strengths up to 770 N/mm².

SPHEROIDAL-GRAPHITE (S.G.) CAST IRON (BS 2789, 'Iron castings with spheroidal or nodular graphite') Ordinary grey cast iron has its free carbon in the form of long graphite flakes, the structure not being conducive to high tensile strength. In S.G. cast iron the graphite is induced to form as spherical or nodular particles, this formation giving higher-strength castings. The effect is brought about by adding small quantities of calcium, cerium, or magnesium to the molten iron, before casting. S.G. iron has a similar tensile strength to that of pearlitic malleable iron.

ALLOY CAST IRON Alloying elements may be added to cast iron in order to improve its mechanical properties, particularly hardness and toughness, and also to improve machinability. *Nickel* and *chromium* are the most common alloying elements, and are often combined together (as in many alloy steels), since they tend to neutralise each other's bad effects. *Molybdenum* and *vanadium* are also used. Below are given three examples of alloy cast irons.

Type	Composition %		Properties
Wear-resistant iron	3·6 C, 0·05 S,	2·8 Si, 0·50 P,	0·6 Mn, 0·17 V	High wear-resistance and long life
Ni–Cr–Mo iron	3·1 C, 0·08 S,	2·1 Si, 0·5 Ni,	0·8 Mn, 0·9 Cr, 0·9 Mo	Hard, strong, and tough
'Ni–hard' iron	3·3 C, 4·5 Ni,	1·1 Si, 1·5 Cr,	0·5 Mn,	Exceptional hardness

Key: C–carbon, Si–silicon, Mn–manganese, S–sulphur, P–phosphorus, V–vanadium, Ni–nickel, Cr–chromium, Mo–molybdenum.

Plain-carbon steel

Like plain cast iron, *plain-carbon steel* is an alloy of ferrite and carbon (plus small controlled amounts of Si, Mn, S, and P), but with a carbon-content of up to 1·25%, all of which is combined with the ferrite. As can be seen from fig. 6.1, plain-carbon steel, again like cast iron, is processed from pig iron. Various steel-making processes are used, but generally the aim in all cases is to reduce the carbon-content in the furnace charge to the desired level. Plain-carbon steels may be grouped into *mild steels*, *medium-carbon steels*, and *high-carbon steels* and are (together with many alloy steels) specified in the comprehensive standard BS 970.

Typical compositions are:

dead-mild steel   0·05 to 0·10%C
mild steel   0·10 to 0·30%C
medium-carbon steel 0·30 to 0·60%C
high-carbon steel   0·60 to 1·25%C

    plus small percentages of Si, Mn, S, and P.

A high-carbon steel containing 0·90–1·25%C is often called *tool steel*, being used for dies, taps, cutters, knives, etc.

All the plain-carbon steels are extensively used in the engineering industry, being cheap and commonplace. The material is available in many raw forms, including hot-rolled (black) or cold-rolled (bright) plate, sheet, or strip (BS 1449); bars, billets, or forgings (BS 970); hot-rolled structural sections such as angle, channel, tee, or joist (BS 4); and drawn wire and drawn tube (BS 2094).

The mechanical properties of plain-carbon steel vary with the carbon-content of the steel and the heat-treatment given to the steel. Figure 6.2 illustrates the mechanical properties of plain-carbon steel in the softest (normalised or annealed) condition.

It can be seen from fig. 6.2 that in general, as the carbon-content of the steel increases, the *tensile strength* and *hardness* increase, while the *ductility* falls. Probably the optimum (most favourable) combination of these mechanical properties occurs with a steel containing 0·87% carbon.

The hardness of a plain-carbon steel can be increased (at the expense of ductility and malleability) by means of heat-treatment. *Hardening* is carried out by heating the steel to a cherry-red colour, followed by quenching in water or oil. The degree of extra hardness achieved depends upon the carbon-content of the steel (the greater the carbon-content, the greater the hardness) and the speed at which cooling takes place (the faster the quench, the greater the hardness); in fact, this heat-treatment will make little difference to the structure of the mild steels. *Tempering* (heating between 200–300 °C) may follow hardening, if required, as this process removes some of the hardness and brittleness while increasing the tensile strength and ductility. Subsequent *annealing* or *normalising* would again bring the steel to its softest and most refined state.

To sum up, the general advantages and limitations of plain-carbon steel are as follows.

ADVANTAGES
a) Cheap
b) Easily worked to various shapes by rolling, forging, or hammering
c) Easily machined
d) Can be welded
e) Various combinations of mechanical properties can be obtained by heat-treatment
f) Available in many different standard forms and sections

LIMITATIONS
a) Suitable only for relatively moderate loading in service
b) The mild steels require special heat-treatment (called case-hardening) to improve their hardness and toughness
c) Poor resistance to corrosion

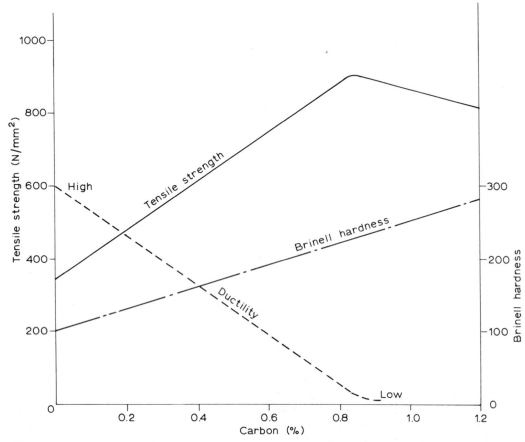

Fig. 6.2 Mechanical properties of plain-carbon steels in normalised condition

An enormous range of components is produced in the engineering industry from plain-carbon steel. Typical examples are nuts, bolts, screws, rivets, wire, nails, shafts, levers, rails, ropes, dies, cutters, tools, bearings, and springs. Many structures and frames of various shapes and sizes are fabricated from plain-carbon structural steel.

6.3 Non-ferrous metals

Metals which do not have iron as their main constituent are called *non-ferrous metals*. This group of materials includes a wide range of metals such as *copper*, *aluminium*, *tin*, *lead*, *zinc*, etc., used on their own as pure metals, or in combination as alloys. Space does not permit consideration of all the non-ferrous alloys, therefore we will confine our discussion to copper and the copper-based alloys, which probably represent the largest group of non-ferrous alloys.

Copper

Most copper originates from ore, the most common ore being copper sulphide. The ore is processed such that pure copper is finally separated from it. Large quantities of copper are used in the electrical engineering industry in the form of wires, cables, etc., as its electrical conductivity (ability to transmit) is outstanding. As its heat conductivity and resistance to corrosion are also excellent, it is frequently used in water-heating equipment, such as pipes and cylinders. However, because it is a soft and weak material, it is not greatly used for general engineering purposes.

Many alloys of copper, which are generally stronger than the parent metal, are widely used for general engineering purposes, and these may be broadly grouped as follows:
a) brasses–alloys of copper and zinc,
b) bronze–an alloy of copper and tin.

Let us consider each in turn.

Brasses

Generally, as the percentage of zinc in brasses is increased, so the ductility decreases and the tensile strength increases. Brasses may be grouped into two types, depending upon the amount of zinc (Zn) alloyed to the copper (Cu).

a) BRASSES CONTAINING UP TO 39% ZINC This group of brasses includes gilding brass (80% Cu, 20% Zn – BS 711), cartridge brass (70% Cu, 30% Zn – BS 267), standard brass (65% Cu, 35% Zn – BS 266 and BS 2870), and common brass (63% Cu, 37% Zn – BS 265). These brasses are most suitable for cold-working, and are available as sheet, strip, foil, tube, wire, and rod. The 70/30 brass has the optimum combination of high ductility and adequate tensile strength [the ratio (yield stress)/(tensile strength) $= \frac{1}{4}$] to allow it to be deep-drawn from sheet blanks upon a press. A deep-drawn cartridge brass may be harder than a mild steel. All the *cold-working brasses* can be annealed, to reduce them to their softest and most ductile state. This group of brasses is not suitable for working to shape at high temperatures. It is used for imitation gold jewellery, ornamental and decorative work, cartridge and shell cases, gas and electric fittings, etc.

b) BRASSES CONTAINING 39% AND MORE ZINC This group of brasses includes muntz metal (60% Cu, 40% Zn – BS 1949). These brasses are most suitable for *hot-working*, such as hot-stamping or forging, but may also be cold-worked to a limited extent. They are available in plate, sheet, rods and bars of various sections, and extruded tube. They machine well, especially in their free-cutting form. Hot-rolled muntz metal is a little stronger than, but not as ductile as, annealed cartridge brass. The brasses containing from 39 to 46% Zn are used for plumbing fittings, condenser tubes, clock and instrument parts, nuts and bolts, etc.

There are a few brasses in use which contain more than 40% Zn, such as 'spelter', which is a term used for brazing brass, this alloy containing 50% Cu and 50% Zn (see section 8.2).

Bronze

Generally, as the percentage of tin in bronze is increased, so the ductility decreases and the tensile strength increases. Bronze is used a good deal for electrical work and bearing applications. A better combination of properties is found in bronze when a further element is alloyed to the copper, in addition to tin. Hence we get phosphor bronze (containing phosphorus), gunmetal (containing zinc), aluminium bronze (containing aluminium), and lead bronze (containing lead).

a) PHOSPHOR BRONZE (BS 369) The composition of this alloy varies depending upon whether it is to be cast to shape or worked in some other way. The addition of phosphorus helps to produce sound castings. Phosphor bronze can be obtained as rod, sheet, or wire, and is occasionally supplied in the form of cast bars or tubes for turning into bearings (see BS 288). The cold-working phosphor bronze which can be drawn into wire or rolled into sheet becomes work-hardened and is very springy. It has excellent resistance to sea-water corrosion, a low coefficient of friction, and great wear-resistance, and is used for steam-turbine blades, valves, bearings, springs, gears, nuts, castings of all shapes, etc.

b) GUNMETAL (BS 1400) Gunmetal has very good casting properties, and the addition of zinc increases the fluidity of the metal as it is poured into the mould in the molten state. It has excellent resistance to corrosion by water or atmosphere, and is therefore widely used for the production of castings for marine purposes. It is not suitable for cold-working, and lead is sometimes added to improve its machinability. Gunmetal is used for pumps, valves, hydraulic-equipment parts, etc.

c) ALUMINIUM BRONZE (DTD 174A, and DTD 197A – note, 'DTD' stands for the 'Directorate of Technical Development', which issues standards similar to those of BSI.) The aluminium bronzes fall into two main groups, containing either approximately 5% or 10% aluminium. The first group contains the cold-working aluminium bronzes, having excellent resistance to corrosion and oxidisation on heating, and a fine gold colour; hence they are much favoured for use as jewellery.

The second group can be hot-worked by forging or casting, having good strength and hardness in this condition. The alloy containing 5% Ni and 5% Fe is a most interesting material, which can be hardened like a plain-carbon steel – i.e. heated to 900°C and quenched, followed by tempering as required – this gives a material hard enough to be used for non-sparking tools, such as files. Aluminium bronze is also used for jewellery (imitation gold), chemical process tubes, gears, worm-wheels, pump rods, marine engine propellers, generator brush holders, etc. Note that aluminium bronze is not considered very suitable for bearings.

d) LEAD BRONZE (DTD 229A) Unlike aluminium–bronze, lead bronze containing up to 35% Pb (lead) is ideal for use as a bearing material. It has a low coefficient of friction and good resistance to wear. It is usually used as a bonded lining inside steel shells; being somewhat plastic in nature, it beds down well to the bearing surface. It is primarily used for heavily loaded bearing bushes.

Exercises 6

1 What are the main characteristics and possible applications of the following: (a) dead-mild steel, (b) mild steel, (c) medium-carbon steel, (d) high-carbon steel?   ULCI

2 a) State the basic composition and two physical or mechanical properties of each of the following materials: (i) mild steel, (ii) cast iron, (iii) hot-working brass.
b) State one disadvantage for each of the above materials.   UEI

3 a) What are the basic constituents of plain-carbon steel?
b) Explain the basic composition differences between grey cast iron and mild steel.
c) Define the following terms used in relation to metal properties: (i) tensile strength, (ii) malleability, (iii) hardness, (iv) machineability.   NCTEC

4 a) How does the carbon-content influence the properties and machining characteristics of a plain-carbon steel?
b) Why is cast iron used for machine-tool beds?
c) What metals are alloyed to make brass?
d) Explain why a large rake angle is suitable when turning mild steel but zero rake is required when turning cast iron.   NCTEC

5 In the assembly shown in fig. 2.18, the shaft is to be made from bright mild-steel bar and the bush from phosphor-bronze cast tube.
a) Explain the properties and characteristics of these materials which make them suitable for (i) the service life of each component in the assembly, (ii) the initial casting process of the bush and the final machining with cutting tools of each of the components.
b) Give the basic composition of each of these materials.   EMEU

6 a) The percentage of carbon in a plain-carbon steel produces variations in the mechanical properties of material. Name these variations.
b) A spindle is required to have hard-wearing qualities. Name two ferrous materials that would be suitable, and state how the hardness could be achieved in each case. Describe a process by which the spindle can be made.   NCTEC

7 Outline the properties, characteristics, and uses of *two* of the following: (a) gunmetal, (b) 70/30 brass, (d) phosphor bronze.   NCTEC

8 Describe briefly the approximate composition and physical properties of the following materials:

(a) gunmetal, (b) cartridge brass.
Give two examples of where each could be used in an engineering component.　　　　　　　UEI

9 a) State a typical composition of (i) medium-carbon steel, (ii) high-carbon steel, (iii) grey cast iron, (iv) bronze, (v) brass. Give two examples of the use of each.
 b) Briefly summarise the contents of BS 970.

10 Cast iron is one of the most widely used metals in engineering. Selecting an example known to you through your own experience, name and sketch a typical grey-cast-iron component. State what were the properties of this material which led to its choice for this particular application.

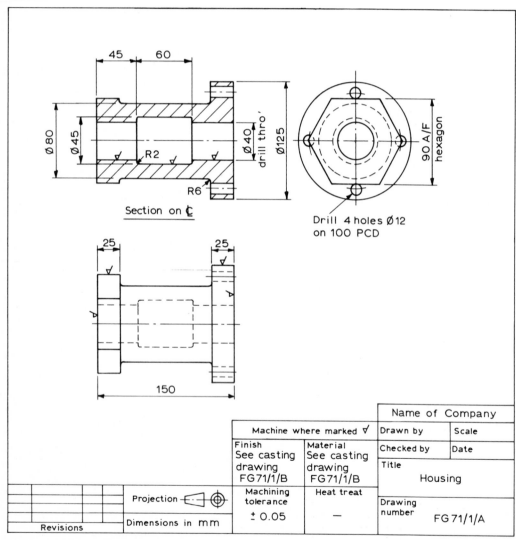

Fig. 7.1 Component single-part drawing

Labels in figure:
45 60
Ø80
Ø45
R2
R6
Ø40 drill thro'
Ø125
90 A/F hexagon
Section on ₵
Drill 4 holes Ø12 on 100 PCD
25   25
150

Title block:
	Name of Company		
Machine where marked √	Drawn by		Scale
Finish See casting drawing FG71/1/B	Material See casting drawing FG71/1/B	Checked by	Date
		Title	Housing
Machining tolerance ± 0.05	Heat treat —	Drawing number	FG 71/1/A

Revisions	Projection ⊏⊏◉	
	Dimensions in mm	

Chapter 7
Sand-casting

1 Drawing castings

In *sand-casting*, the oldest and best known of the casting processes, molten metal is poured into a sand mould which is either 'green' (i.e. no steps are taken to dry out the mould after it has been made) or 'dry' (i.e. the mould is dried, and hence hardened, prior to pouring the metal–this being more suitable for large castings). Many moulds are now made from sand (sodium silicate) which is dried and hardened by the application of carbon dioxide (CO_2), but the majority of sand-castings are still produced in green moulds.

Patterns made of wood, plastics, or metal are used to form the mould, which lasts only while one casting is produced. Castings of mass varying from 30 g up to hundreds of tonnes can be made by this process, with an average dimensional tolerance of approximately 4 mm. The surface finish is also only moderately good, and therefore much machining is usually carried out on castings in order to bring them to their final shape and finish. All the basic materials mentioned in Chapter 6 (cast iron, plain-carbon steel, brass, and bronze) may be cast to shapes both plain and intricate.

It was stated in section 5.4 that each individual part to be produced by an engineering company should be drawn separately on a single-part drawing, complete with machining instructions. If such a part were needed to be cast to shape before final machining (see fig. 5.6(a), for example), then a copy of the single-part drawing would be required by the pattern-maker. A pattern-maker is a highly skilled tradesman who is capable of producing a pattern which will satisfy the requirements of both the foundry moulder (who must be able to make a satisfactory mould from the pattern) and the machinist (who must have sufficient metal left in the right place on the casting for machining, this being known as the 'machining allowance').

In some cases, particularly where intricate castings are required, it is necessary to produce single-part drawings both of the required part and of the casting from which the finished part will evolve. The casting single-part drawing will include details of necessary machining allowances, draft (taper), which allows the pattern to be easily withdrawn from the mould without damaging it, and any holes which must be produced in the casting by coring. Note that holes of diameter less than 25 mm cannot usually be cored in castings, and must be machined from the solid metal. The finished casting will be smaller than the pattern, due to contraction upon cooling; the pattern-maker compensates for this contraction by making the pattern larger than the casting. This dimensional addition is known as the 'contraction allowance', the following values giving a general idea of how much different metals contract.

Metal	Contraction allowance (%)
Plain steel	1·6
Cast iron	1 to 2
Brass	2
Bronze	1·3 to 2

Figure 7.1 shows a single-part drawing of a simple component which is to be machined to final shape from a casting. Figure 7.2 shows the single-part drawing of the casting for the same component. The following points should be noticed when comparing fig. 7.2 with fig. 7.1.

a) Draft or taper is shown on some casting faces, to facilitate withdrawal of the pattern from the mould; this draft must taper in the direction in which the pattern will be withdrawn (see section 7.2).

b) The casting drawing shows a plain hole through the casting; this will be produced by coring at the moulding stage (see section 7.2). (In practice, this would be rather a difficult coring job, as the length is several times greater than the diameter.) This cored hole will then be machined to its final shape. The four 12 mm diameter holes must be produced by drilling from the solid metal, as they are too small to be cored.

c) A machining allowance of 10 mm on diameter and length has been added to those parts of the casting which are to be machined to size.

d) The hexagon feature is large enough to be cast to shape (prior to machining), instead of being cast to a circular shape. This latter alternative would leave an excessive amount of iron to be removed by machining.

7.2 Production of simple castings

In this section, a description of the process of sand-casting will be given, using the casting shown in fig. 7.2 as an example to illustrate the process. It will be necessary to make a wooden pattern, around which the mould will be formed, and a wooden core-box, in which the core will be formed. The process is outlined step by step, as follows.

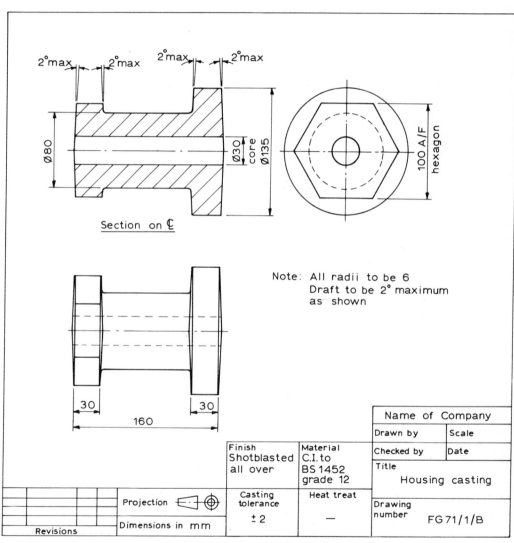

Section on ₵

Note: All radii to be 6
Draft to be 2° maximum as shown

		Name of Company	
		Drawn by	Scale
Finish	Material	Checked by	Date
Shotblasted all over	C.I. to BS 1452 grade 12	Title	
		Housing casting	
	Heat treat		
Projection ⊏⊐ ⊕	Casting tolerance ± 2	—	Drawing number FG 71/1/B
Revisions	Dimensions in mm		

Fig. 7.2 Casting single-part drawing

79

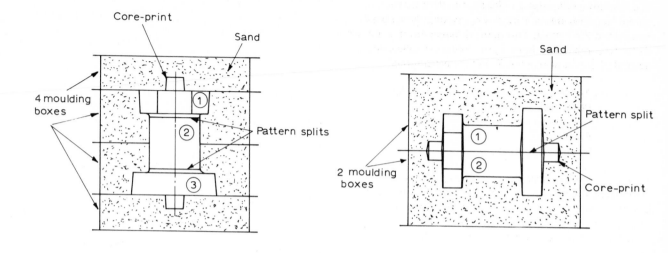

(a) Pattern in 3 parts
Axis vertical

(b) Pattern in 2 parts
Axis horizontal

Fig. 7.3 Moulding method

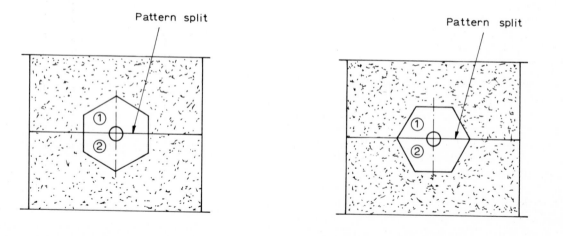

(a) Pattern split across the flats

(b) Pattern split across the corners

Fig. 7.4 Moulding method

1) Manufacture of pattern

The wooden *pattern* must be made in two parts (sometimes more), so that it can easily be removed from the mould. The plane along which the pattern is split usually represents the parting line of the mould; hence it can be seen that the pattern design is determined entirely by the method used for moulding. In our example, it will be better if the part is cast with the axis horizontal rather than vertical, because the former method is simpler –requiring only two parts for the pattern. Figure 7.3 shows the difference.

Figure 7.3(b) shows a method which is clearly less complicated and expensive that that shown at fig. 7.3(a) and is therefore the better method. Note that the chosen moulding method dictates the direction of the draft on the pattern.

Figure 7.3 shows core-prints, which are an integral part of the pattern. These projections on the pattern leave impressions, or prints, which locate and support the core in the mould. There is one further point to consider about the pattern, concerning the hexagonal feature. There are two alternative ways of splitting the pattern along the axis when viewed from the end elevation. This point is shown in fig. 7.4, and it is clear that fig. 7.4(b) is the better alternative when considering pattern withdrawal from the mould.

The pattern halves are dowelled so that they fit accurately together as one in the mould. Finished patterns are smoothed and then painted to a colour code, viz. orange (as-cast surface), yellow (machined surface), and black (core-prints).

2) Manufacture of core

A split core-box (dowelled for location, like a split pattern) is used to mould the *core* to shape. This box is filled with special core sand, opened, and the core is carefully withdrawn. The core is then baked in an oven until hard and porous.

A sketch of the core-box is shown in fig. 7.5. In a complicated casting requiring several cores, it would be necessary to make a core-box for the production of each.

3) Preparation of mould

a) The pattern half with dowel holes (not pegs) is placed in a moulding-box or flask (called a *drag*). The pattern half is sprinkled with fine dry parting sand, then the drag is completely filled up with moulding sand. This is firmly rammed around the pattern half, before being levelled off

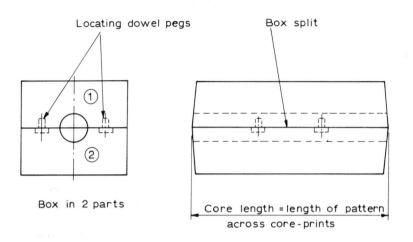

Fig. 7.5 Core-box

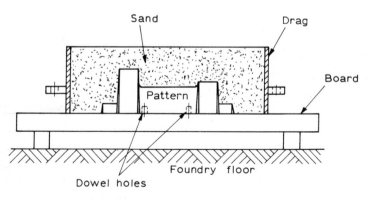

Fig. 7.6 Drag rammed up

81

or *strickled*. This stage is shown completed in fig. 7.6.

The drag is then turned over, and a further layer of parting sand is applied across the flat face of the moulding sand. This parting sand helps prevent the pattern sticking in the mould, or the mould halves sticking together.

b) The top half of the pattern is assembled to the lower half, still contained in the drag. The top moulding-flask (called a *cope*) is fitted to the drag, and is in turn filled and rammed with sand as before. At this stage, cavities are prepared in the cope for the down-gate (also called the 'pouring gate' or 'sprue') and the riser, as shown in fig. 7.7.

The down-gate is used as a spout into which is poured the molten metal. When the mould is filled, excess metal will fill up the riser, giving a head of metal through which gases can escape, hence ensuring a solid casting. The riser also acts as a reservoir of metal, to counteract shrinkage during the cooling stage.

c) The moulding-flasks are separated, and the cope is turned over so that the pattern halves can be removed from the boxes. Runners or channels from the down-gate to the mould cavity are now formed in the sand, these runners being required to allow the molten metal to run and enter the mould at the best access point.

d) The core is now fitted into the core-prints in the drag, and the cope is assembled to the drag, thus closing the mould. The completed flask is clamped, and sometimes a pouring cup is fitted above the down-gate, to ease pouring. The clamping must be sufficiently firm to prevent the cope lifting off the drag as molten metal fills the mould cavity. Figure 7.8 shows the completed mould, ready for pouring.

4) Removal of casting from mould

After cooling, the casting is removed from the separated moulding-flasks. The core is broken up and removed from the casting. The sprue and riser are then cut off, and the casting is fettled or dressed by grinding and shot-blasting.

The process of sand-casting has several shortcomings in that the structure of the finished casting is frequently poor, being coarse-grained and sometimes having blowholes. The process is slow and the finish is rough. Superior castings may be produced by using metal dies instead of sand moulds, but the cost of production is higher, and the process is suitable only for certain non-ferrous metals; therefore, one still finds sand-moulding in frequent use, despite its shortcomings, because of its versatility and cheapness.

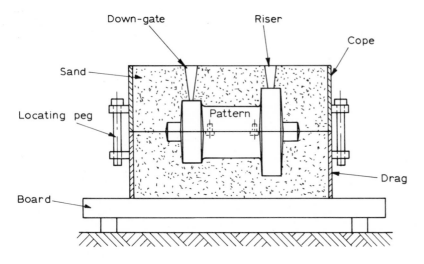

Fig. 7.7 Cope and drag rammed up

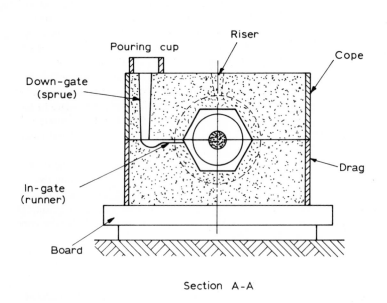

Section A-A

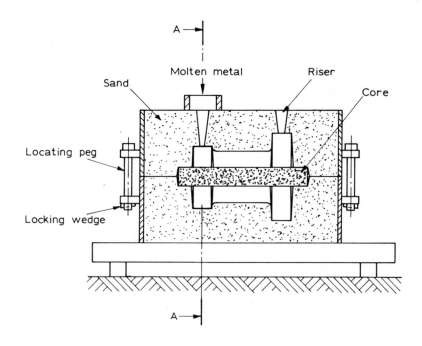

Fig. 7.8 Completed mould, ready for pouring

Exercises 7

1 The component shown in fig. 7.9 is to be produced by sand-casting in grey cast iron.

a) Outline the essential stages employed for producing the casting.

b) Sketch the pattern required, showing a suitable machining allowance for the bore and end faces. WJEC

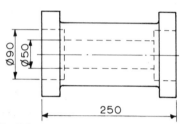

Fig. 7.9

2 Describe with the aid of sketches the method by which a simple iron casting is made with the aid of a loose pattern and moulding-boxes. ULCI

3 a) Two stages of contraction are involved in the cooling of a casting in a sand mould: (i) the liquid stage, and (ii) the solid stage. How are these compensated for?

b) The quadrant support bracket shown in figs 4.19 and 7.10 is to be cast in a sand mould. The 19 mm diameter hole and the slot are to be machined at a later stage, but the 32 mm diameter must be cored.

Sketch in good proportion a section through the mould, clearly indicating (i) the position of the mould joint; (ii) the core, showing how it is located; and (iii) the down-gate or runner. NCTEC

4 a) Draw a typical casting, and show on your drawing (i) machining allowance, (ii) draft angle, (iii) blend radii.

b) Why are the above needed when casting? UEI

5 The fixture body in fig. 5.18 may be produced by casting in a sand mould.

a) With the aid of sketches, explain how the sand mould would be prepared and how the subsequent casting process would be carried out. Indicate, in particular, how the molten metal is introduced into the mould, and state what safety precautions are required.

b) List the advantages and disadvantages of this method of producing components. EMEU

6 a) Briefly describe the stages in the preparation of a sand mould for producing the casting shown in fig. 3.28.

b) Make an undimensioned sketch of (i) the pattern, (ii) the core-box. NCTEC

7 a) A casting is said to be porous and to contain blowholes. Explain what these defects are, and give possible causes commonly encountered.

b) Describe fully how a mould would be prepared for sand-casting the crank-arm shown in fig. 7.11. NCTEC

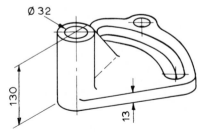

Fig. 7.10

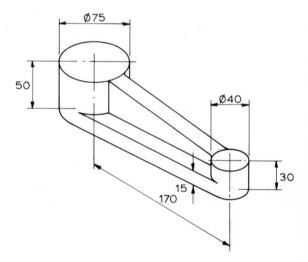

Fig. 7.11

8 The location bracket shown in fig. 7.12 is to be manufactured as a sand-casting instead of being prefabricated. Redraw this detail as a casting, fully dimensioned, and state below your drawing suitable values for machining allowances, blend radii, and draft angles.   UEI

9 a) Explain why a wooden pattern is larger than a cast-iron component produced from it.

b) Why may another pattern be required if the same component is cast in aluminium?   UEI

10 You are required to sand-cast a disc 75 mm in diameter × 75 mm wide with a 30 mm diameter cored hole in the centre. Explain with the aid of sketches how you would prepare the mould ready for pouring the metal.

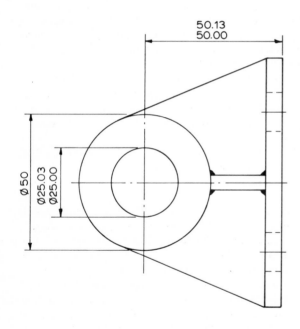

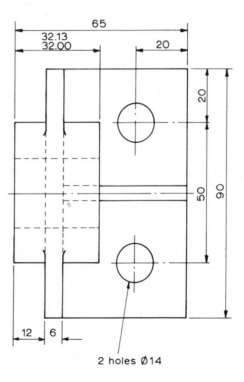

Fig. 7.12

Chapter 8 Fabrication & welding

1 Drawing fabricated components

A *fabricated* part is made up of several component parts (cut from plate, sheet, or structural sections) which are then fastened together by riveting, bolting, or welding. Of these joining processes, welding can be said to give the best results, for the following reasons:
a) it is quick, simple, and cheap;
b) it saves on material, giving a lighter construction;
c) the finished article looks 'one piece'.

It should be mentioned here, though, that adhesives (particularly those of resin composition) are increasingly being used to join together parts made from various materials.

One will often find that the processes of (a) fabricating and joining by welding and (b) casting are direct alternatives for the production of a particular part to be made from metal; the part functioning equally well in service when produced by either method. The following factors must be taken into account when making a choice.
a) Cost – castings are cheaper in large-batch production, but dearer than fabrications when only a few are required.
b) Appearance – a cast shape is more attractive than a fabricated one, but this must be counted as a matter of opinion.
c) Application – most metals can be sand-cast to shape, but some metals are not easily welded; for example, the weldability of a dead-mild steel is high, but the weldability of a high-carbon steel is low.
d) Machinability – generally, a casting is more easily machined than is a fabrication.
e) Strength – both cast and fabricated parts can be designed to withstand operating forces equally well.

Fabrication work can be roughly divided into thin-plate work (up to 3 mm thick) and thick-plate work (over 3 mm thick). The methods used in these two fields of fabrication engineering may be different in that much *thin metal sheet* can be shaped cold, but *thick metal plate* must often be heated before bending to shape. Welding and riveting can be used for both, but soldering or brazing is most successful when applied to thin metal sheet.

When producing drawings of fabricated and welded parts in a drawing office, it is advisable to use standard symbols to represent welding and welded joints, as shown in BS 499, 'Welding terms and symbols'. A brief selection of joints and welding symbols for both thin and thick plate is shown in fig. 8.1.

The most common types of welds used are *butt* and *fillet*, the latter being a little weaker than the former. It can be seen from fig. 8.1 that the preparation of two plates which are to be welded when butted together (edge to edge) is different when using thick plate rather than thin. It is necessary to bevel the edges of thick plate, in order to provide an enlarged groove to hold the weld metal. A lap (overlapped) joint is an alternative way of joining two plates using fillet welds, but the finished result is not neat.

Figure 8.1 shows symbols to be used upon a drawing to represent welds, as recommended in BS 499. The use of these symbols saves time, as the draughtsman is cutting down on unnecessary detailing (see section 3.4), and the welding requirements specified on the drawing will be clear and unambiguous. Figure 8.2 shows a selection of welding symbols used as drawing instructions, with their interpretation.

It can be seen from fig. 8.2 that a welding symbol can be used to denote weld-type, position, and size. In fig. 8.2(a), the weld symbol should be placed above the reference line (not below, as shown) if a single weld is required on the opposite side of the joint to that shown.

Figure 8.3 shows a detail drawing of a fabricated bracket which is to be produced by welding separate pieces together. Figure 8.4 shows a sketch of the same part as it would be designed if it were to be produced as a one-piece sand-casting.

The welding symbol used in fig. 8.3 indicates that fillet welding is required all round to a size of 5 mm. The bracket shown is fabricated from four separate parts. If one such bracket is being made by a skilled welder/fabricator, then the parts may be cut (using an oxygen-acetylene gas-welding torch) or sawn to shape (occasionally machined before welding), and he will take his sizes directly from the drawing. On the other hand, if the brackets are being produced in large quantities, then separate single-part drawings of the four component parts will be required. These component parts will be made in a fabrication department, being finally welded together by production operatives using fixtures to position and hold the parts. Any machining required will be carried out last, after the fabricating and welding process is completed.

When comparing the *fabricated bracket* (fig. 8.3) with the *cast bracket* (fig. 8.4), note that the latter has a uniform thickness for the base and ribs. This in turn leads to uniform cooling, hence avoiding the possibility of stress

Sketch of joint	Type of joint	Type of weld	Plate thickness	Plate edge preparation	Weld symbol
	Butt	Butt	Up to 3mm	Square	
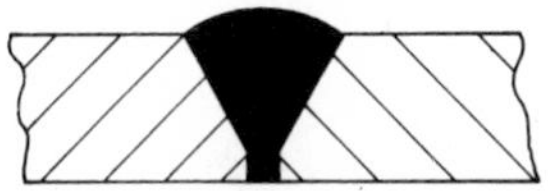	Single-vee butt	Butt	Over 5mm	Bevel	
	Double-vee butt	Butt	Over 12mm	Double bevel	
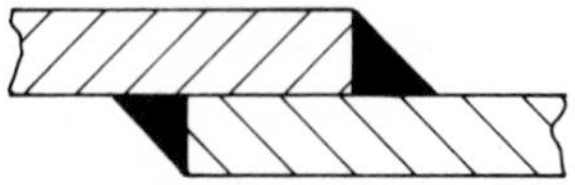	Lap	Fillet	Over 5mm	Square	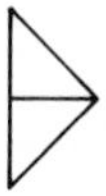
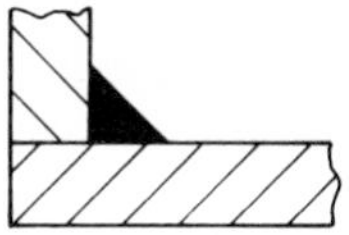	Corner	Fillet	Over 5mm	Square	
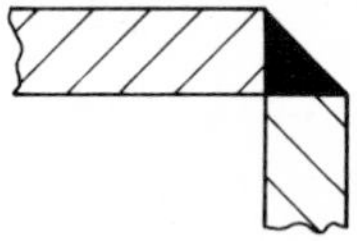	Corner	Fillet	Over 5mm	Square	
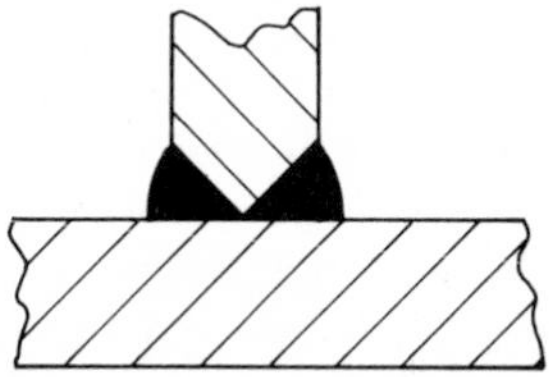	Tee	Butt	Over 12mm	Double bevel	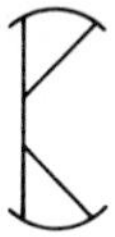

Fig. 8.1 Welding joints and symbols

Drawing instruction	Practical interpretation

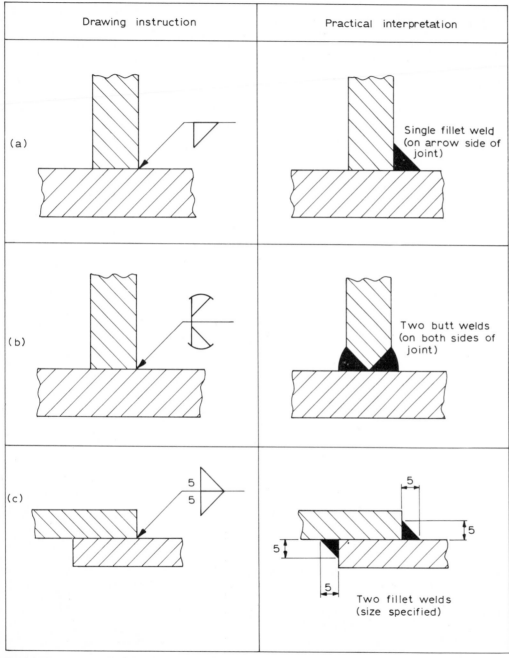

(a) Single fillet weld (on arrow side of joint)

(b) Two butt welds (on both sides of joint)

(c) 5 / 5 — Two fillet welds (size specified)

Fig. 8.2 Interpretation of welding symbols

cracks occurring. This design feature is not necessary on the fabricated bracket, which is still sufficiently strong if the vertical ribs are made half the base thickness.

8.2 Joining by soldering and brazing

The processes of *soldering* and *brazing* are similar in that complete fusion (melting and blending) takes place at the surfaces of the pieces of metal being joined. The depth of the fusion or blending varies with the type of materials and the temperatures being used, and surface fusion is relatively shallow. Let us consider each process in turn.

Soldering

Two metal parts may be joined together by running a thin layer of molten solder between them, the solder being applied to the joint by means of a heated soldering-iron. The molten solder is drawn into the narrow joint space between the parts by *capillary action* (This is the name given to the phenomenon where liquids rise against the force of gravity; it can occur only in fine spaces or bores.) A low melting-point alloy of lead and tin in wire or stick form is used for *soft soldering* (as it is called). Since the strength of soft solder is low, the process is used only on joints which are to be lightly loaded, such as thin sheet-metal parts. The process can be used on iron, steel, and the copper-based alloys, but aluminium is difficult to solder, due to its high oxidation rate.

A flux (aid to fluidity) must be used when soldering. An active flux such as zinc-chloride liquid ('killed spirits') is common, this cleaning the metal by dissolving away any oxide film which might be present. Such a flux is corrosive, and must be cleaned off after soldering.

Brazing (see BS 1723, 'Brazing')

As with soldering, two metal parts may be joined together by running a thin layer of molten brazing solder (called 'spelter') between them. This process is sometimes called *hard soldering*, particularly when a silver solder (29% copper, 10% zinc, 61% silver) is used. Brazing spelter (50% Cu, 50% Zn) is obtainable in stick, strip, and granular form, and has a higher melting-temperature and strength than soft solder. Brazing can therefore be used to make stronger joints, and is suitable for use on most of the ferrous and non-ferrous metals.

The parts to be brazed must be chemically clean, and borax (hydrated sodium borate, obtained by evaporating

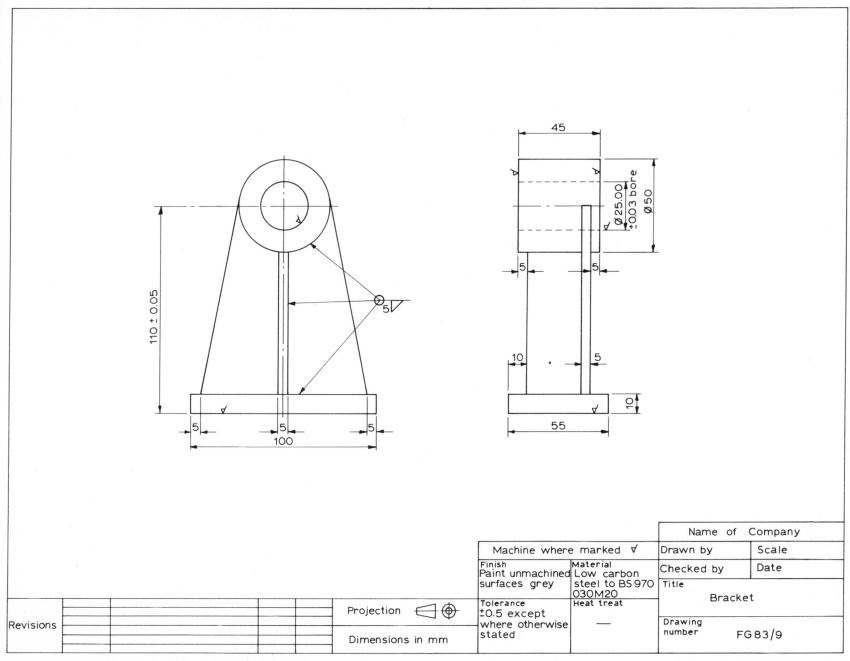

Fig. 8.3 Fabricated bracket

89

alkaline lake-water, and sometimes used as an antiseptic or as a hard-water softener) in paste or powder form is used as a flux. After the application of flux, the metal parts are assembled with spelter between them, in strip or grain form, and then heated by means of a gas torch, or in a furnace, until the melting-temperature of the spelter is reached and fusion takes place. If a spelter stick is used, it must now be applied to the heated joint, so that it runs between the parts, due to capillary action. This process is often carried out in a hearth.

8.3 Joining by welding

Welding processes can be divided into two groups, viz. *fusion-welding* (where separate parts are jointed together by the application of heat and a filler-rod at the joint) and *pressure-welding* (where parts are joined by the application of heat and pressure). We shall here have space to consider only the first process of the two. Total fusion occurs in fusion-welding (as opposed to surface fusion in soldering and brazing) because the joint edges of the metal parts being joined become molten and fuse together as one.

Welding is a large subject, and there is a variety of

fusion-welding methods used, but we will consider only the two best-known and most widely used processes, which are *gas-welding* and *arc-welding*. Before examining the processes individually, we will compare them by listing their advantages and limitations.

a) The temperature of an electric arc is much higher than the temperature of a gas flame; therefore arc-welding melts the joint faces practically instantaneously – hence the operation is quicker and cheaper.

b) Gas-welding is unsuitable for plates above 20 mm thick (because of reasons given in (a) above), and pre-heating of the work may be necessary.

c) Gas-welding is the cheaper process when used on plates' less than 6 mm thick.

d) Gas-welding can be used on most metals, and is particularly favoured for non-ferrous materials.

e) Arc-welding is simpler and faster, requiring only a suitable power-supply. Gas-welding, with the gas-bottles mounted on a trolley, is portable.

Gas-welding (see BS 693 'Oxy-acetylene welding of mild steel'; BS 1453, 'Filler rods for gas-welding')

The most commonly used gases are oxygen and acetylene, which are usually supplied in pressure vessels called gas-cylinders. These cylinders are painted to a colour code (see BS 349) so that the contents can easily be identified, the code being black–oxygen, maroon–acetylene. The gases from the cylinders are passed through reducing valves to the tip of the torch (or blowpipe), where they are burned in approximately equal proportions, giving a flame temperature of about 3000°C. Gas-welding equipment must always be treated with respect, as in certain circumstances acetylene is explosive (when a flame is applied to it under pressure), as is oxygen when used in an oily atmosphere (such as an old, dirty garage floor-pit). Also, a high-temperature flame from a hand-held torch is dangerous when used carelessly. Welding should never be attempted before adequate instruction has been received from an expert.

The torch is held in one hand (usually the right), and the torch flame is applied to the joint so that the metal edges melt. At the same time, a metal filler-rod, held in the other hand, is applied to the welding zone so that the end melts into the molten pool. As flame and rod are slowly advanced along the joint, a weld of any required size is built up between the metal parts being joined. The process is illustrated in fig. 8.5.

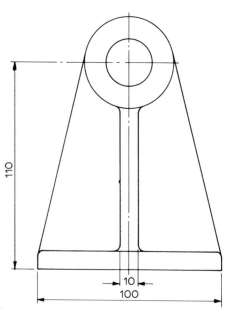

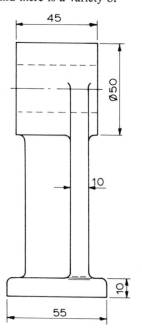

Fig. 8.4 Cast bracket

Figure 8.5 depicts a single-vee butt joint being welded, with the torch and rod simultaneously moving from right to left; although sometimes the reverse may be used. Goggles having tinted glass must be worn to protect the eyes during this operation. A more difficult job than the one shown might necessitate the component parts being held together by clamping.

Many common type of filler-rods are available; mild-steel rods, for example, being used when welding mild-steel parts. Cast iron, in particular, is better welded using a bronze filler-rod and a special flux. This particular form of the process is known as *bronze-welding*, and is akin to brazing.

Arc-welding (see BS 638, 'Arc-welding plant, equipment, and accessories')

In this process, the heat necessary to melt the metal joint edges is obtained from an electric arc struck between the electrode (filler-rod) and the work, giving a temperature approaching 4000°C in the welding zone. A current of low voltage and high amperage is required for this process, and is provided by an a.c. transformer welding-set or a d.c. generator set, the latter being more expensive but superior for use with cast iron or the non-ferrous metals. The filler-rod is coated with a suitable flux, which forms a fusible substance with any oxide present (leaving a slag which must be chipped off after welding) and also assists in stabilising the arc. Filler-rods are available in all the weldable metals.

The *electric-arc process* is easier than gas-welding, as the welding rod is clamped in a holder which is held in one hand. After the arc has been struck, the rod is slowly moved along the joint, the free hand being required only to hold a protective face-shield (unless a head-fitting-type shield is used). The process is shown in fig. 8.6.

Figure 8.6 shows a single-vee butt joint being welded, with the filler-rod being moved from left to right; this is the most convenient set-up for a right-handed operator. Hot metal tends to spatter during electric-welding, and protective gloves and clothing are even more essential than in gas-welding.

8.4 Joining by riveting

At one time, *riveting* was a widely used method of permanently joining two pieces of metal together; today it has been replaced by welding in many applications. This is because welding is faster, cheaper, and simpler.

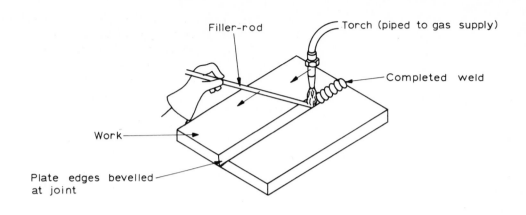

Fig. 8.5 Oxy-acetylene welding

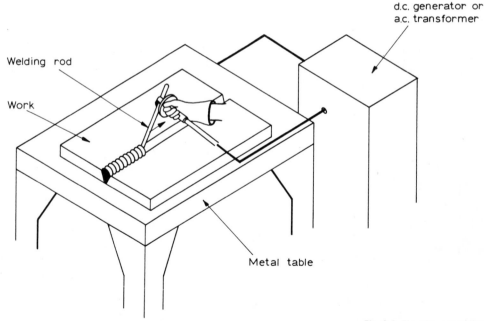

Fig. 8.6 Electric-arc welding

91

The aim in welding is to use simple butt joints wherever possible, instead of the more expensive lap joint. In riveting, lap joints have to be used, or the more complex butt joint with cover-plate(s) – shown in fig. 8.7.

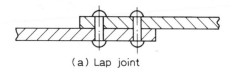

(a) Lap joint

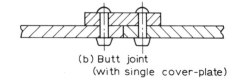

(b) Butt joint
(with single cover-plate)

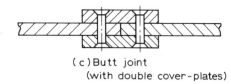

(c) Butt joint
(with double cover-plates)

Fig. 8.7 Riveted joints

The butt joints in fig. 8.7 should be compared with the simpler form of welded joint shown in fig. 8.1. The most common forms of rivets are shown in fig. 8.7, these being:
a) cup or snap head – riveted over to the same shape at the opposite end;
b) pan head – riveted over to a snap head at the opposite end;
c) countersunk head – riveted over to the same shape at the opposite end, hence leaving the rivets flush with the work surface. Countersunk holes are necessary in this case.

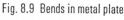

Fig. 8.8 Riveting operation

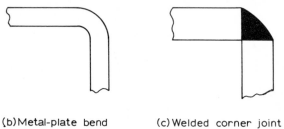

(a) Sheet-metal bend   (b) Metal-plate bend   (c) Welded corner joint

Fig. 8.9 Bends in metal plate

Rivets are made of a ductile material such as mild steel, copper, brass, or duralumin – an alloy of aluminium. Depending upon the size of the rivet, they may be riveted over (as shown in fig. 8.8) hot or cold.

After suitable sized holed have been drilled or punched in the plates, the rivets are positioned one by one in the holes and supported by a tool, as shown in fig. 8.8, before being closed or clinched up under pressure. A snap and a hand hammer may be used for clinching, or with large quantities an automatic gun (either pneumatically or hydraulically operated) may be used. A special mechanical gun is used to close up 'pop' rivets, which are a small hollow rivet used in thin sheet-metal work. With countersunk rivets, flat-headed tools are used to squeeze the rivets closed.

8.5 Development of fabricated shapes
As we have seen so far in this chapter, much engineering work consists of building up parts into a *fabricated whole* by joining using bolting, riveting, welding, soldering, or brazing. Much thin-gauge sheet metal is formed to shape by bending a suitably shaped flat piece into the 'three-dimensional' final object. With thicker plate, an allowance has to be made when calculating the original length required, as the thicker material is difficult to bend without a corner radius. This is demonstrated in fig. 8.9.

If a simple right-angle shape is required, as shown in fig. 8.9, it may be assumed that a sharp bend can be produced in thin sheet metal [fig. 8.9(a)] but a bend radius is necessary in thick metal plate [fig. 8.9(b)], this sometimes requiring red heat to facilitate bending. A possible alternative is a welded assembly [fig. 8.9(c)]. Yet again it can be seen that the designer must be fully aware of the advantages and limitations of alternative engineering processes, making the function and the cost of the final product his criteria. In this book we shall have room only to introduce the subject of development, and we will assume that thin sheet metal is being used; therefore no 'bend allowances' will be considered.

The process of determining the *original (developed) shape* of a flat metal piece which is to be bent into the *finished article* is called '*development*'. For convenience, we will call the flat developed shape the 'pattern' and the bent final shape the 'product'. The setting out of a pattern is a geometric exercise, and a skilled sheet-metal worker will deduce and mark out the pattern shape directly onto the sheet of metal he will cut and bend to shape. In the case of large-quantity production, the draughtsman will draw and dimension the pattern shape. From this single-part drawing, a template will be produced (in wood or metal), which in turn will be used for scribing around onto the metal sheets from which the products are to be cut and bent. Figure 8.10 shows a simple example of the development of an open square metal tray.

Figure 8.10(a) shows the developed pattern shape, including the corner tabs which are to be folded inside the tray. The fold or bend lines are shown dotted (for the purposes of illustration only), and the scrap material left from the rectangular sheet after the pattern has been cut out is shown shaded. In the case of thin sheet metal, the corner joints will be completed by soldering, the tabs giving a greater soldering area and hence a stronger joint. With thick plate, the corners may be joined by welding, the tabs then not being necessary. In practice, the top edges of the tray would be folded over a little way, to form a rounded 'safe edge' in preference to the raw sharp edge left by cut sheet metal.

Regular straight patterns of the type shown in fig. 8.10 are relatively easy to visualise and set out, but now consider the more difficult exercise of making a pattern to produce an open-bottomed square pyramid (assuming the joint is to be edge welded, and no tab is required). The pattern and product for the pyramid are shown in fig. 8.11.

The actual plan and elevation of the pyramid are shown in fig. 8.11. The difficulty here is that line O_1A on the pattern must be equal to the true length of the sloping line on the product from the base corner to the apex. This true length (known as the *slant height*) cannot be scaled from either the plan or the elevation, but could be calculated by the use of trigonometry. However, in practice it is easier to set out the pattern using geometric methods. Referring to fig. 8.11, the procedure is as follows.

a) Draw the actual plan and elevation to scale.

b) With centre O_{11}, draw an arc of radius equal to OA, hence giving the length $O_{11}A$ (equal to the plan length OA).

c) With centre O_1, draw a part circle of radius equal to O_1A. (Note–O_1A is the true length of the slant height of the pyramid.)

d) With centre A on the circle, draw an arc of radius equal to AB. Join line AB. Repeat with equal lengths BC, CD, and DA.

The pattern is completed by joining the lines O_1A. Basically, the pyramid pattern was set out using the

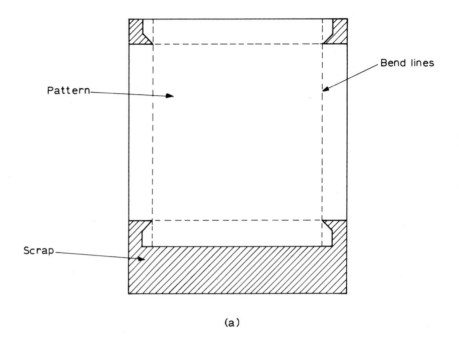

Pattern

Bend lines

Scrap

(a)

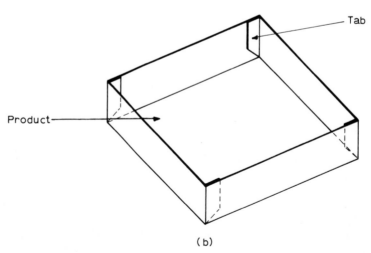

Tab

Product

(b)

Fig. 8.10 Development of a tray

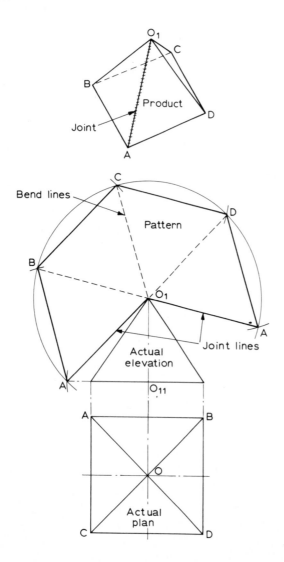

Fig. 8.11 Development of a pyramid

radial-line construction method. This method may be applied to any product shape which can be generated by a straight line, one end of which remains stationary while the other end traces out any path. (Apply this definition to the pyramid example–it helps if one can visualise objects in three dimensions!) Let us consider a more difficult example using the same method.

Radial-line construction

A little thought will show that the pattern of a cone will be the sector of a circle of radius equal to the cone slant height and of arc-length equal to the cone base circumference. The development of the frustum of a cone (i.e. a truncated cone) is a little more difficult, and is shown in fig. 8.12. The top and bottom of the product are open, giving a 'choirboy-collar' shape.

Comparison of figs 8.11 and 8.12 will show the similar construction methods; but, in the case of a cone, the true length of the slant height (O_1A) can be derived directly from the actual elevation. Referring to fig. 8.12, the procedure is as follows.

a) Draw the actual plan and elevation to scale.
b) Divide the circumference of the plan (base) circle into 12 equal parts, and letter the points A, B, C, etc.
c) Project vertical lines upwards from these points A, B, C, etc. to intersect the cone base in the elevation.
d) Join these intersecting points in the elevation to the cone apex O_1.
e) With centre O_1 and radius O_1A, draw a part circle circumference.
f) With centre A on this circle, draw an arc of radius equal to AB. Repeat with equal arcs BC, CD, etc. to the last one, MA. (At this stage, the development of a full cone is complete.)
g) Project horizontal lines from the intersection points on the truncation line across to the slant-height line O_1A.
h) With centre O_1, draw part circles of radius O_1a, O_1b, O_1c, etc.
j) Where these circles and the radial lines emanating from O_1 intersect, mark points a, b, c, etc., round to the last point, a.

The pattern is completed by drawing a smooth curve through points a, b, c, etc., and joining the lines aA.

Parallel-line construction

This method is used instead of the radial-line method when the product shape can be generated by a line which moves parallel to the product axis. We will consider the development of a truncated cylinder as an example of the use of this method, this being shown in fig. 8.13. Note that two such products joined together would give an 'elbow', as shown dotted.

Referring to fig. 8.13, the procedure is as follows.
a) Draw the actual plan and elevation to scale.
b) Divide the circumférence of the plan circle into 12 equal parts, and letter the points A, B, C, etc.
c) Project vertical lines upwards from these points A, B, C, etc. to the truncation line.
d) Draw the enclosing rectangle of the pattern by projecting horizontally across from the elevation, making the rectangle length equal to the circumference of the plan circle (i.e. pattern length AA = plan-circle circumference).
e) Project horizontal lines from the intersection points on the truncation line across through the rectangle.
f) Divide the pattern length AA into 12 equal parts: AB, BC, CD, etc. Erect perpendiculars from points A, B, C, etc.
g) Where vertical and horizontal ordinates intersect, mark points a, b, c, etc., through to the last point, a.

The pattern is completed by drawing a smooth curve through points a, b, c, etc., and joining the lines aA.

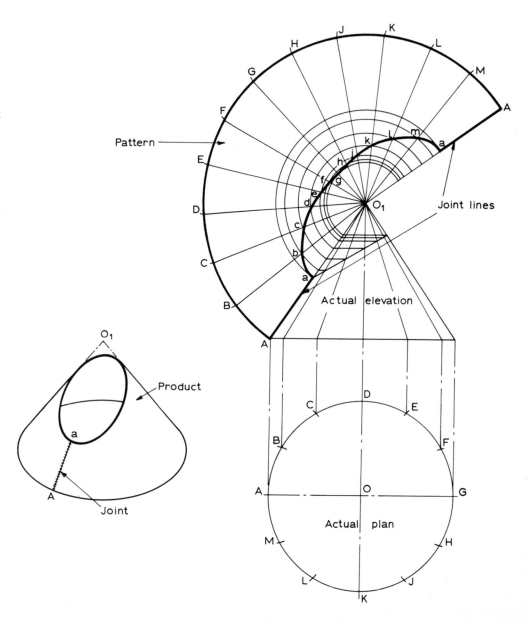

Fig. 8.12 Development of the frustum of a cone

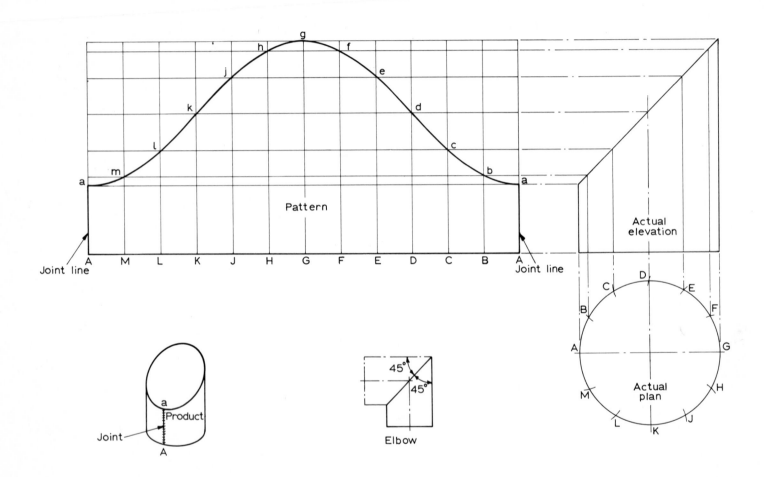

Fig. 8.13 Development of a truncated cylinder

Triangulation construction

This is an alternative constructional method which may be used for the purposes of development. The product will be formed from the pattern by folding along the sides of a series of triangles. In the case of curved components, folding will replace rolling, and the resulting shape is most attractive. One has to imagine the surface of each product, then, to consist of a series of triangles.

As an example, let us consider the development of a 'square-to-round' hood, such a shape being in common use in many engineering works. The method of pattern construction is shown in fig. 8.14.

Referring to fig. 8.14, the procedure is as follows.

a) Draw the actual plan and elevation to scale.

b) Divide the plan circle into 12 equal parts, and letter the points A, B, C, etc.

c) Number the square corners and joint positions 1 to 5.

d) From the corners of the square, draw straight lines to these points.

e) To one side of the elevation, draw true-length diagrams as shown, setting off lengths 1A, 5M, and 5A (from the plan) as shown. Join O5 (perpendicular), OA, and OM on this diagram. These are the true lengths of the slant heights of the triangles emanating from each corner. (See the similar problem in the development of a pyramid, fig. 8.11.) Note–the true lengths of 5M, 5L, 2B, 2C, 3E, 3F, 4J, and 4H are the same; likewise the true lengths of 5A, 5K, 2A, 2D, 3D, 3G, 4G, and 4K are the same.

f) In any suitable position, draw the straight line 1A on the right of the pattern equal to its true length O(A) on the true-length diagram.

g) From point 1, erect perpendicular 1–2 equal in length to 5–1 (plan).

h) With a dotted line, join 2A. (This should equal length OA on the true-length diagram.)

j) With centre A on the pattern, draw an arc equal in radius to AB (plan).

k) With centre 2 on the pattern, draw an intersecting arc equal in radius to the true length of 2B, which is OM on the true-length diagram.

e) With a dotted line, join 2B.

Continue the development in a similar manner until all points A, B, C, . . ., A and 1, 2, 3, . . ., 1 are plotted. Join all the outside points with straight lines and all the inside points with a smooth curve which, when cut and joined, forms a circle.

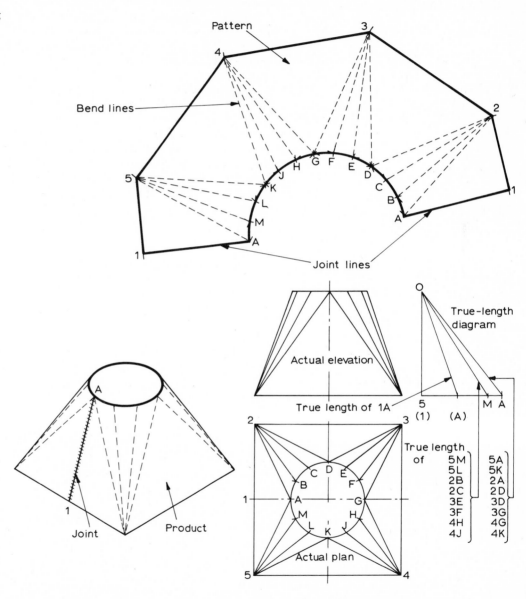

Fig. 8.14 Development of square-to-round hood

The astute reader will have noticed in two examples that arc distances AB, BC, CD, etc., stepped off the plan circles for the development (see figs 8.12 and 8.14), are in fact chordal distances. This leads to a little loss in accuracy, the development being only approximately correct, but it is often good enough in most fabrication work. Where one wishes to be exact, the true arc lengths must be calculated as a fraction of the circle circumference ($\frac{1}{12}$ in the examples shown).

A great range of development work exists where fabricated pipe and duct assemblies are used. These consist basically of intersecting pipes of any section joined at any angle. This necessitates a developed hole (the shape of which is known as the 'curve of interpenetration') being cut in one pipe, this being covered by a developed shape on the end of the intersecting pipe. Radial- or parallel-line construction methods are used; unfortunately, space does not permit the showing of further examples.

Exercises 8

1 Figure 8.15 shows a hopper to a square-section feed-tube. Both hopper and feed-tube are to be made in one piece from 3 mm plate.

a) Draw three views of the assembly, in orthographic projection, showing a suitable weld size and symbol in accordance with BS 499.

b) Develop the pattern for the hopper. Make no allowance for the metal thickness and joint. Leave in all the construction and true-length lines.   NCTEC

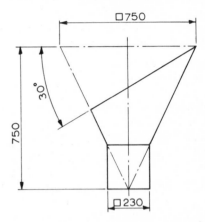

Fig. 8.15

2 What are the main considerations to be taken into account when deciding if a component is to be prefabricated or cast?   UEI

3 a) With the aid of sketches, explain one use for each of the following types of jointing metal: (i) plumber's solder, (ii) tinman's solder, (iii) hard solder. State the composition of the solder in each case, and give the reasons for variations.

Explain the different fluxing agents required for each of the solders, and state why fluxes are necessary when making joints.

b) Explain the method of making any one of the soldered joints you have shown, with particular reference to the heat source and its control, the final joint strength, and the economics of the process.   EMEU

4 Figure 8.16 shows a bracket made by welding together standard sections.

a) Sketch two orthographic views, and add the welding instructions by means of standard symbols to BS 499. Explain and show where 'edge preparation' is required.

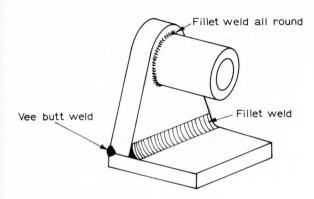

Fillet weld all round

Vee butt weld

Fillet weld

Fig. 8.16 Welded bracket

b) Explain the term 'fusion-welding', and say why this method of joining may be selected for the component, in preference to brazing.

c) State which method of fusion-welding would be used for this component and which welding method would be more suitable for sheet-metal components.　EMEU

5 Figure 8.17 shows three truncated cylinders forming a three-piece elbow. Draw, scale full-size, the developed shape of each piece.　NCTEC

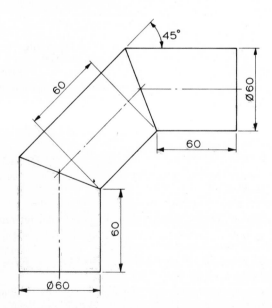

45°

60

Ø60

60

60

Ø60

Fig. 8.17 Three-piece elbow

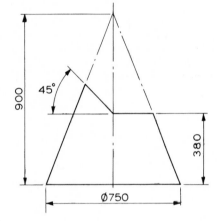

900

45°

380

Ø750

Fig. 8.18

6 Figure 8.18 shows a cone made from sheet metal cut by an offset plane. Using any suitable scale, (a) develop the full pattern for the cone, making no allowance for joint or material thickness; (b) sketch the type of watertight joint that may be used if the cone is to be made from copper sheet, and show what modifications would be made to the developed pattern. (c) State how the joint would be made if the cone were made from 3mm plate, and show the appropriate symbols that would be used on the drawing to indicate the precise details.　NCTEC

7 a) Define the following terms: (i) weld, (ii) parent metal, (iii) filler metal.

b) What is meant by 'gas-welding'?

c) Use clear diagrams to explain the essential differences between leftward and rightward gas-welding.　NCTEC

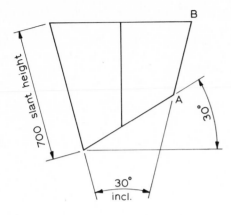

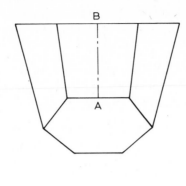

10 a) Distinguish between a cast product and a fabricated product.

b) Discuss the factors that should be considered in deciding upon the method of manufacture to be adopted for a new product.

c) Explain how contraction is compensated for when casting a component which will be machined.　NCTEC

11 The tray shown in fig. 8.20 is to be made of tinplate 0·5 mm thick. Allowing 6 mm for the overlaps, develop the pattern and indicate the bend lines.

Describe the method of completion, mentioning the equipment and materials to be used.

Discuss the design of this tray with regard to safety.

WJEC

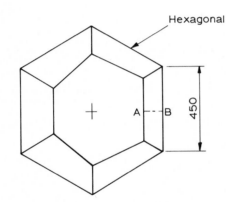

Make joint along AB

Fig. 8.19

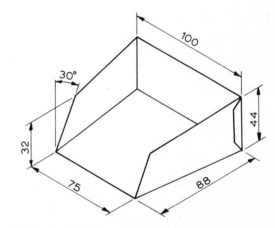

Fig. 8.20

8 Figure 8.19 shows a special hopper manufactured from light-gauge tinplate.

a) Develop a full pattern for the hopper, to a scale of 1:10. No allowance need be made for joint or material thickness.

b) Make a large-scale detailed sketch of a suitable continuous watertight joint, to run along the line AB.　NCTEC

9 a) Explain the essential differences between a welded joint and a brazed joint.

b) Is a brazed joint stronger than a soft-soldered joint when joining together two pieces of mild steel? Give reasons justifying the answer.　UEI

12 a) (i) Make a neat sketch of two parts which you would rivet together. Give two examples of the practical application of riveting. (ii) What are the advantages, if any, of hot-riveting?

b) (i) Make neat sketches of two examples of metal components joined by soft solder, and give reasons for the choice of joining method. (ii) Under what circumstances would hard soldering or brazing be chosen rather than soft soldering? (iii) What is the purpose of a flux in soft soldering?　ULCI

9.1 Properties of pipes

Pipes (or tubes, which are the same thing) are manufactured from various materials, both metallic and non-metallic, and are used extensively for industrial and domestic applications. They are the vehicles for transmitting air, water, gas, oil, and steam from the source or main to any required point (and for housing electric wire, in the case of conduit tube–see Chapter 10). Very broadly speaking, water-pipe circuits are usually installed by a plumber, and other pipe circuits by a pipe-fitter; however, the basic principles underlying all *pipework* are similar in whatever field they are applied. Pipework installations are considered to be part of the factory services, and as such come under the jurisdiction of the plant or works engineer.

Pipe is nominally classified by the bore or inside diameter, but it is convenient to keep outside diameters constant (to facilitate threading) and vary bore sizes to give different wall thicknesses. The bore must be of a suitable dimension to accommodate the volume of the substance being transmitted. The pipe-wall thickness or gauge must be sufficiently thick for any given material to withstand the required working pressure without failure (and the subsequent risk of accident). Note that all the substances mentioned earlier are transmitted at varying pressures, but a very general classification might be town gas and water–low pressure (0–3 bars), air and steam–medium pressure (3–20 bars), and oil in hydraulic pipelines–high pressure (20–400 bars). There are, of course, exceptions to these broad groupings. (Note: 1 bar = 10^5 pascals = 10^5 N/m².)

Many British Standards are in existence to cover all applications (e.g. BS 61, 'Heavy-gauge copper tubes for general purposes', or BS 3284, 'Polythene pipes for cold-water services'), but they are far too numerous to be listed here. However, we will briefly consider the range of materials from which pipes are made, together with some typical applications.

COPPER (see BS 61, BS 659, BS 1306, and BS 2017) Copper pipe is manufactured in solid drawn form, having a smooth and attractive finish which does not rust. It is used extensively for water, town gas, and air, and for hydraulic-fluid branch lines (from the main service line) at low and medium pressures.

STEEL (see BS 1387, BS 3601, and BS 3602) Plain-carbon steel pipe, of solid-drawn or welded-seam construction, is available for all purposes. Stainless-steel pipe, which costs considerably more than plain-steel pipe, is frequently used for conveying corrosive chemicals.

CAST IRON (see BS 1211 and BS 2035) Spun cast-iron pipe is manufactured by a centrifugal casting technique in which a metal mould rotates. It is mainly used in the low-pressure range for water or town-gas mains.

PLASTICS (see BS 1972, BS 3284, BS 3505, and BS 3796) Attractive and cheap plastics pipe materials are available, including PVC (polyvinyl chloride), PE (polyethylene), PP (polypropylene), and ABS (acrylonitrile butadiene styrene). They are resistant to corrosion by chemicals or the atmosphere, but are limited to low pressures and temperatures. They are now used extensively for cold-water applications.

OTHER MATERIALS Other materials such as rubber, lead, brass, aluminium, and wrought iron are used less frequently than those listed above.

There are two further points to consider here, both concerned with safety. Great care must be taken over the choice of pipe for conveying hydraulic fluids (pressure risk), or town gas (ignition risk) for example. Expert advice should be taken when in doubt about the type of pipe to use. Also, a colour code exists (see BS 1710) by means of which the contents of a pipe can be identified; examples are drinking water–aircraft blue, compressed air (up to 14 bars pressure)–white, town gas–canary yellow, hydraulic power–salmon pink (ground colour)/sea green (bands). Service pipes which represent a particular hazard are striped black and yellow.

9.2 Pipe fittings (see BS 864, BS 1256, BS 1740, BS 2051, BS 4622, and BS 4772)

It would be impracticable to make a pipe run or circuit from one piece of piping; therefore standard fittings are used for a variety of purposes, these fittings being joined to straight or curved pieces of pipe so that the whole forms one complete, leak-proof assembly. A representative selection of the more common fittings in use is shown in fig. 9.1.

All the fittings shown in fig. 9.1 may be obtained plain, screwed, or flanged, depending upon the joining method to be used (see section 9.3), these alternatives being shown with

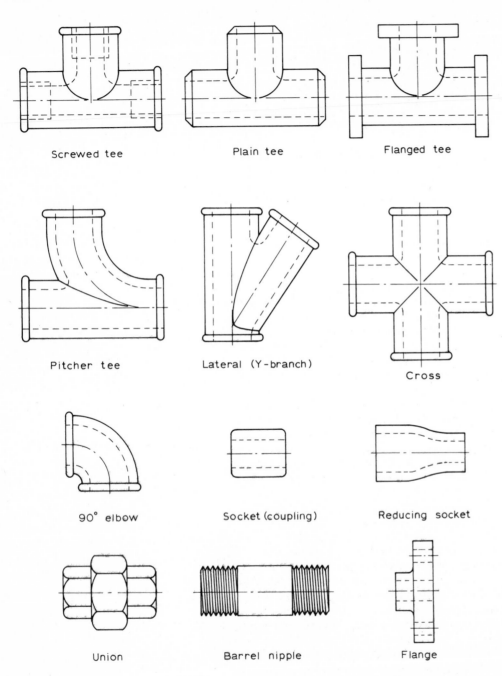

Screwed tee   Plain tee   Flanged tee

Pitcher tee   Lateral (Y-branch)   Cross

90° elbow   Socket (coupling)   Reducing socket

Union   Barrel nipple   Flange

Fig. 9.1 Pipe fittings

102

respect to the tee fitting. The fittings are sized according to the nominal bore of the pipe for which they are to be used, and screwed fittings are not usually recommended for pipe larger than 40mm diameter bore.

Fittings are fabricated, forged, or cast from a variety of materials, including steel, malleable iron, bronze, copper, and plastics. Clips may also be mentioned here, these being shaped collars which encircle the pipe and enable it to be fixed to a wall or column, etc. A variety of clips exists for different applications.

Pipe-joining methods

A great many methods are in existence for joining two or more pipe lengths together. The choice of joint depends upon the pipe material, size, and the purpose for which it is being used. Examples of a few of the more common joints in use are given below.

a) FLANGED JOINTS (see BS 1560, BS 4622, and BS 4772)
A *flange* is shown in fig. 9.1, and this may be cast integrally, welded, or screwed onto the ends of a pipe. Flanges on mating pipes are then bolted together face to face. *Joint-rings* (gaskets) are usually fitted between the faces, to provide a good seal. Figure 9.2 shows a typical flanged joint.

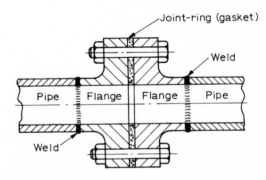

Fig. 9.2 Flanged joint

The type of flange illustrated in fig. 9.2 is called a welding neck flange (as opposed to a flange which is slipped over the outside of the pipe before welding.) The flanged faces are machined, and the joint-ring is compressed between them. This type of joint would be suitable for a steel pipe which was to be used on a hydraulic installation at 17 bars maximum pressure.

Joint-rings are available in different sections (such as corrugated, oval, or 'O'-section, for example) and different materials (such as compressed asbestos fibre, rubber, or plastics, for example), depending upon the operating conditions.

b) SCREWED JOINTS (see BS 21) The British Standard Pipe (BSP) thread specified in BS 21 is the preferred series of threads for *screwed pipe joints*, and has now been adopted as the International Pipe Thread, (ISO recommendation R7). External threads are designated male threads; these may be either parallel or taper. Two methods of making screwed joints for pressure purposes are commonly used:

   i) taper male fitting with parallel thread, or
   ii) taper male fitting with taper thread.

Ideally, ferrous fittings should be used with ferrous pipes wherever possible, and likewise for non-ferrous materials— this avoids galvanic attack between copper and iron. Screwed plastics fittings may also be used with either metal or plastics pipe.

A great variety of screwed joints is in existence (excluding screwed flanges, mentioned earlier), but in general they can be classified into three groups:

   i) joining by socket,
   ii) joining by union,
   iii) joining by nipple.

Given the additional alternative of right- or left-hand threads, any pipe layout should be possible. Figure 9.3 shows one example from each group.

The joints in fig. 9.3(a) would be suitable for metal pipes used for a low-pressure application. Left-hand and right-hand threaded sockets could be used at opposite ends of the pipe run; this configuration allows a pipe length to be dismantled without disturbing any other part of the layout.

Union joints [fig. 9.3(b)] are used particularly in steam circuits, being simple to fit and less liable to leakage than a socket joint. The jointing faces are usually conical, as shown, or hemispherical.

Figure 9.3(c) shows a useful application of a simple screwed nipple where lack of room necessitates the direct coupling together of two fittings. In practice, the fittings are screwed right up, no part of the nipple being visible.

In all cases, gaskets are necessary for pressure-tight and leak-proof joints.

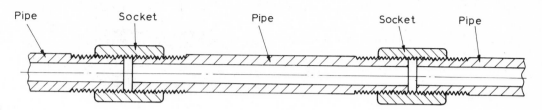

(a) Socket joint

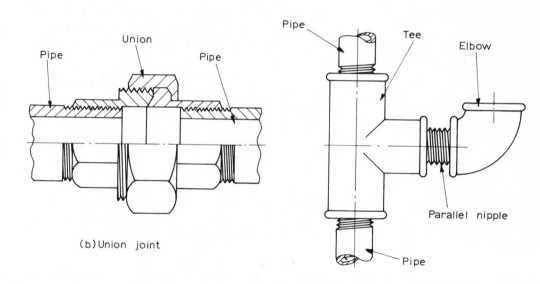

(b) Union joint

(c) Nipple joint (partially closed)

c) COMPRESSION JOINTS (see BS 864 and BS 2051) Strictly speaking, all joints that incorporate a joint-ring which is compressed at the assembly stage are compression joints; however we are referring here to joints which are made up by using compression fittings. The use of these has increased considerably in recent years, as they can be used with lighter gauge tube than is possible with screwed fittings, and they are also very quick and easy to assemble. Figure 9.4 shows a typical joint made up with a compression fitting.

Fig. 9.3 Screwed joints

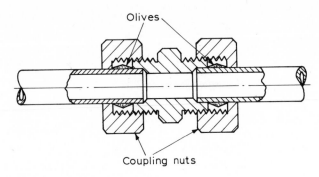

Fig. 9.4 Compression joint

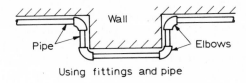

Using fittings and pipe

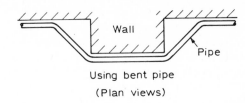

Using bent pipe
(Plan views)

Fig. 9.5 Pipe pass-over bend

The joint shown at fig. 9.4 requires only that the pipe ends are cut square. As the coupling nuts are tightened up, the 'olive' (ring) is compressed, and grips the outside of the pipe. These joints are utilised extensively on thin-gauge copper tubes such as are used in central-heating installations.

d) OTHER JOINTS Other pipe joints are used in addition to those mentioned above. Many pipes are married together using plain fittings which are connected by welding, soldering, or brazing. Capillary-soldered joints ('sweated' joints), for example, are often used on copper tubes, and larger copper tubes may be brazed or bronze welded together. Further variations may be seen in plastics-pipe installations. Here, pipes are often cold welded using a solvent to plasticise the two pipe ends being coupled together. Many plastics pipes are simply fitted together on a spigot-and-socket location: when the pipes are pushed together, an 'O'-ring completes the seal.

9.4 Methods of bending pipes

When a pipe circuit has to pass around a building, the pipe will have to be routed over and around obstructions such as columns, roof trusses, other pipes, etc. It is sometimes more convenient to *bend* a pipe length around an obstruction, rather than to use fittings coupled to short pipe lengths, as shown in fig. 9.5.

The offset or double-bend shown in fig. 9.5 saves on fittings, has no leakage problems, and provides a better flow-path. It is advisable to use a template as a pattern for bending, which may be carried out hot or cold. Copper tubing, in particular, bends easily, but all metal pipe can be bent provided the correct technique is used. To prevent the pipe walls collapsing at the bend, it will be necessary to

support them internally during bending. This is done by 'loading' or filling the tube with a convenient material, which is then emptied out of the bore after bending. The loading materials commonly used are (a) sand, (b) lead or a low melting-point metallic alloy, and (c) steel spring, for light-gauge tube.

Small or light-gauge pipe may be bent manually, using a spring, with one end supported in a block hole or held in lead clamps in a vice. The pipe may be pulled around a former if required. Greasing of the bore will help release the spring.

Larger pipe is best bent (hot or cold, depending upon size) on a bending machine, which may be manually or mechanically operated. Ideally, hot bending should be carried out at one heating, but this is not always possible.

9.5 Drawing pipework circuits

When pipework circuits are being designed to carry different services around a factory, one may be faced with a complex problem in setting the requirements out upon a drawing. Nowadays, much use is made of three-dimensional models, which are built to scale to represent the complete installation. Once practical routes for the various pipes have been established with the aid of the model, the information can be recorded upon a drawing. (Ideally, as mentioned earlier, the different pipe circuits, once installed, should be painted in their appropriate colours for identification). Less complex pipe circuits can be drawn up in plan, side, and/or end elevations, without the aid of models.

In any event, it is advisable to use the graphical symbols

for pipes outlined in BS 1553 part 1 when setting out pipe layouts on a drawing. The justification for using these symbols is exactly as explained earlier in sections 3.4 and 8.1. The selection of symbols shown in fig. 9.6 is simple and self-explanatory; the Standard itself shows many more, of course.

We will complete the chapter by showing an example of part of a relatively simple pipe circuit as portrayed symbolically upon a drawing, this being shown in fig. 9.7. This may be compared with a drawing of the piping as it actually looks, this being shown in fig. 9.8. As an introduction to the subject, it will be sufficient to show it in only one elevation, and without a parts list, although in practice this would be an incomplete specification. The various fittings are shown labelled in fig. 9.8, this being done for clarity.

A complete pipe circuit in the style of fig. 9.7, drawn to scale, would enable the correct quantities of fittings and pipe to be ascertained before installation. This process is known as 'taking-off quantities'.

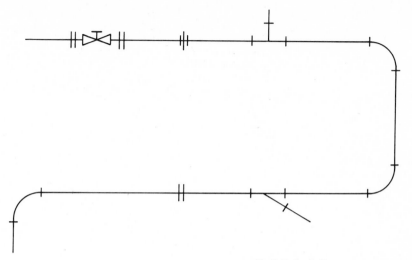

Fig. 9.7 Symbolic representation of a pipe circuit

Feature	Symbol
Pipe	or
Flanged and bolted joint	
Screwed (male and female) joint	
Butt-welded joint	
Hand-operated valve	
Power-operated valve	(Take broken line to control unit)
Safety or relief valve	

Fig. 9.6 Graphical symbols for pipes and valves

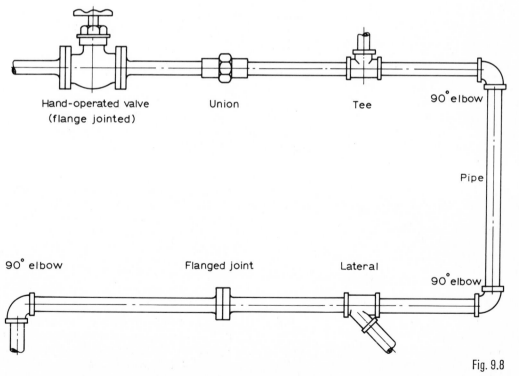

Fig. 9.8

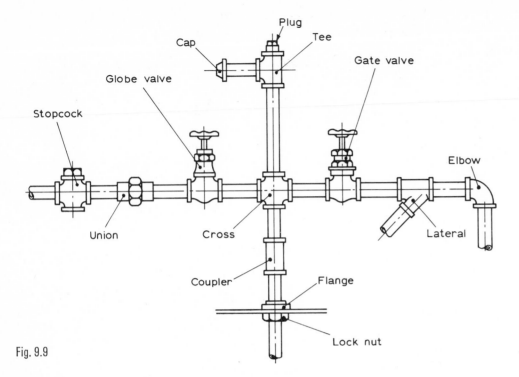

Fig. 9.9

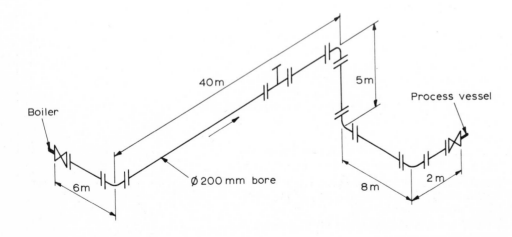

Fig. 9.10

106

Exercises 9

1 a) State the circumstances in which a union joint would be used when connecting two pieces of straight iron pipe.

b) Make a sketch showing a section through a union joint suitable for connecting two straight pieces of iron pipe, showing clearly the principle and method of location and securing.

c) Figure 9.9 shows a double-line layout of a pipework circuit. Redraw the layout using a simple single-line system. NCTEC

2 Figure 9.10 shows the arrangement of a steam pipeline connecting a boiler to a steam-heated process vessel.

a) Explain the advantages of this diagrammatic form of drawing.

b) Make a list, in general terms, of the items of plant in the line, showing the symbol in each case for reference.

c) Explain, with the aid of a sketch, the method of connecting any one of the items in the pipeline, and indicate how leakage is prevented at the joint.

d) Explain, with the aid of sketches, a compression joint suitable for connecting a small-bore PVC tube to laboratory equipment. EMEU

3 a) What is the most significant fault likely to occur when bending copper pipe?

State what precautions may be taken to avoid this fault when producing a right-angle bend in (i) 6 mm diameter copper pipe, (ii) 12 mm diameter copper pipe, (iii) 50 mm diameter copper pipe.

b) The front and side elevation of a header tank that is to supply water to three mixing taps is shown in fig. 9.11. Make a simple line diagram of the pipework and fittings required to complete the circuit. The piping system should be so arranged that each mixing tank may be removed from the circuit for repair without disturbing the function of the others. Indicate on the drawing the fittings required. NCTEC

4 a) Make sketches to show any three of the following terms used when referring to the joining of tubes and tube fittings: (i) screwed tubes with socket union, (ii) a tube with a screwed tee-joint, (iii) a screwed elbow joint, (iv) a flanged tube with holes 'off centres'.

b) Explain the methods used and the precautions required when bending (i) 25 mm bore, medium-gauge, mild-steel tubing, (ii) 12 mm diameter, thin-walled copper tubing. EMEU

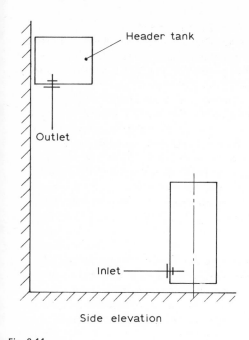

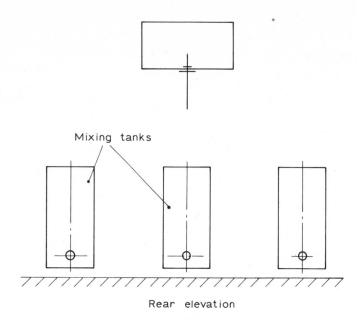

Side elevation

Rear elevation

Fig. 9.11

5 a) The following is an instruction for a piping circuit. Commencing in the top left-hand corner with a 'tee', complete the circuit using a simple single-line representation.

Plug one tee outlet. From one outlet, start the circuit and include the following fittings in order, ending by closing the circuit at the remaining tee outlet: (i) globe valve, (ii) union, (iii) 90° elbow, (iv) check valve, (v) 90° elbow, (vi) 45° lateral, (vii) gate valve, (viii) 90° elbow, (ix) stopcock. The free end of the circuit from the 45° lateral should be capped.

b) Draw a sectional sketch through each of the following straight couplers used for connecting two pieces of straight copper pipe: (i) a compression-type coupler, (ii) a soldered-type coupler. NCTEC

6 a) Make a single-line schematic diagram of a pipe run which will include the following: (i) stopcock, (ii) pressure gauge, (iii) temperature gauge, (iv) 90° elbow, (v) gate valve, (vi) tee-junction with one side capped.

b) Draw a sectional sketch through a screwed and bolted flange joint suitable for hydraulic pipes. NCTEC

7 a) Make a neat sketch of a compression-type joint suitable for use on copper pipe.

b) State two advantages that this type of joint has over a soldered joint.

8 Plastics pipes are now specified for rainwater services for the greater proportion of all new houses. Suggest reasons why you think plastics materials have replaced other materials previously used for this purpose.

9 a) Both flanged and screwed joints are used extensively in industrial pipework. Sketch one joint of each type, and state its advantages and disadvantages.

b) Describe the function of a joint-ring in a flanged joint, and name two materials from which joint-rings are commonly made.

10 A cold-water tap is required on the outside wall of a kitchen, the pipe to it being connected into the existing vertical water-pipe leading up to the sink-tap inside the kitchen. 18 mm diameter copper tube and compression fittings are to be used.

Draw (a) a symbolic diagram and (b) an actual drawing of a possible pipe circuit, showing all the fittings used.

Chapter 10
Electrical assemblies

Fig. 10.1 Graphical symbols for general electrical purposes

Feature	Symbol
Conductors (not connected)	
Conductors (connected)	
Contact (separable)	○
Contact (not separable)	●
Plug	
Socket	
Fuse	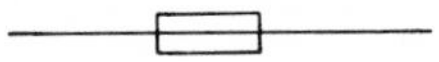
Switch (general symbol)	 or  (Both drawn open)
Push-button switch	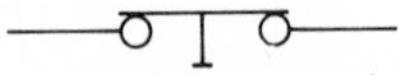 (Drawn closed)
Filament lamp (general symbol)	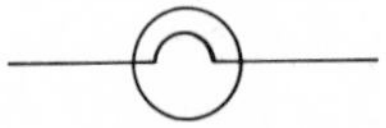
Electric bell	
Battery	 + ‖‖‖ −
Ground or earth (general symbol)	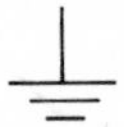
Three-phase electric motor	 M

1 Symbolic notation of wiring circuits

Diagrams, using symbolic notation, are preferred to actual drawings when setting out electrical installation requirements. The principles underlying the procedure are similar to those described for pipe circuits in section 9.5. Again, it is advisable to consult the appropriate British Standard for the preferred symbols, the standard in this case being BS 3939, 'Graphical symbols for electrical power, telecommunications, and electronic diagrams', which incorporates the older standards BS 108 and BS 530. A representative selection of graphical symbols for general electrical purposes is shown in fig. 10.1.

The diagrams most frequently used for electrical purposes are (a) block diagrams, (b) circuit diagrams, and (c) wiring diagrams.

a) BLOCK DIAGRAMS Specific symbols (as shown in fig. 10.1) are not used in this type of diagram, the various pieces of equipment being represented by a simple square or rectangle, as the name implies. Each rectangle or *block* is labelled to indicate its purpose. No attempt is made to draw the installation to scale. Figure 10.2 shows a block diagram.

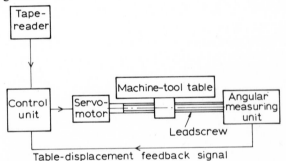

Fig. 10.2 Block diagram

The example shown in fig. 10.2 is of a punched-tape-controlled machine tool. Clearly the diagram is not to scale, and no indication is given of the method of connection, only of the order of connection; hence the advantages of the block diagram can be seen–viz. that a very complex installation can be simply represented.

To briefly explain fig. 10.2, a series of electrical signals is transmitted by the control unit to the leadscrew (driving) servo-motor, the duration of the signals or impulses depending upon the pattern of punched holes on the tape being fed through the tape-reader. As the leadscrew rotates, the table will move the controlled distance. Simultaneously, the angular measuring unit on the leadscrew end records the rotation of the leadscrew (and hence the table movement), and sends return (feedback) signals to the control unit. Providing the input and feedback signals are in accord, the table movement continues as desired; if not, the control unit stops.

b) CIRCUIT DIAGRAMS A circuit diagram makes use of the graphical symbols shown in fig. 10.1 in order to represent the various pieces of equipment in an electrical assembly connected together in a circuit. As opposed to a block diagram, the intention is to show how the circuit functions, but not how it is wired up. An example of a simple circuit diagram is shown in fig. 10.3.

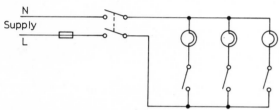

Fig. 10.3 Circuit diagram

The diagram shown in fig. 10.3 is explicit, the symbols requiring no labelling on the diagram. The circuit shown is a master-control lighting circuit, three lamps being shown connected in parallel to an a.c. single-phase supply (if one lamp fails, the others will still light up when switched on). The live wire (L) and neutral (N) for earth wire are marked accordingly. The circuit is used where individual lamps are each required to be under the control of individual single-pole switches, all of which can be over-ridden by a master double-pole switch.

c) WIRING DIAGRAMS These are similar to circuit diagrams, and the same graphical symbols are used in their construction. The main difference is that wiring diagrams are more practical in nature, often showing terminal connections and wire colours so that mistakes are avoided when connecting up equipment.

Figure 10.4 shows a small part of a typical wiring diagram for a motor car; again it will be noted that no attempt is made at accurate scale and location for the

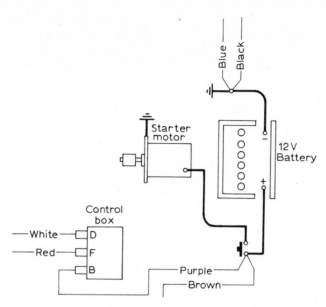

Fig. 10.4 Wiring diagram (part)

various items of equipment. Wiring diagrams are often found inside the cover of electric-motor terminal boxes.

In addition to the three basic types of diagrams shown above, *layout drawings* may be used, indicating the correct position of electrical equipment, and also the required conduit (tubing) runs (see section 10.3). The principles involved are similar to those discussed in section 9.5 for pipework circuits. This type of drawing would be used by electricians, say, installing the electrical services in a new building. Like pipework circuit drawings, electrical layout drawings enable the correct quantities of fittings, conduit tubing (and the wire which runs through the tubing), etc. to be determined before installation.

Safety
Electricity can be lethal, and, because of the special dangers involved in mishandling electrical equipment, we might briefly consider the regulations which exist to ensure safe practices.

a) ELECTRICITY (FACTORIES ACT) SPECIAL REGULATIONS, 1908 and 1944. It is legally binding that all electrical equipment installed in factories and workshops is to the standard stated in these regulations.

b) THE ELECTRICITY SUPPLY REGULATIONS (issued by the Department of Trade and Industry) These regulations were drawn up with the object of 'securing the safety of the Public, and for insuring a proper and sufficient supply of electrical energy'. These regulations do not apply to premises which come under the jurisdiction of the Factories Act Regulations (a) above.

c) THE REGULATIONS FOR THE ELECTRICAL EQUIPMENT OF BUILDINGS (issued by the Institution of Electrical Engineers) The 'IEE reg's', as they are known in the trade, cover the whole field of electrical installation work and are the electrician's bible. It can be said that, if these non-statutory regulations are complied with, then the work should generally satisfy all other regulations.

d) THE CENTRAL ELECTRICITY GENERATING BOARD publish useful booklets on the subject of safety.

Space does not permit consideration of the above regulations, but we might usefully refer to reg. 28 from the Factories Act Regulations before we close this section. This regulation states, 'Any electrical work that would be *dangerous* without sufficient technical knowledge or experience may only be carried out by an authorised person, or a competent person over the age of 18 under his immediate supervision.'

10.2 Conductors and insulators
Certain materials have in them free electrons moving randomly in all directions at speeds of many thousands of metres per hour. If one such material were connected across a battery, then an electric field would exist. The function of the battery is to keep the flow of electrons (or flow of electricity) going. Any material having free electrons is known as a *conductor*. Metals possess large numbers of free electrons, and are therefore good conductors of electricity (and also good conductors of heat).

Copper is a particularly good conductor, and is also easily drawn into wire or cable; it is therefore widely used for both indoor and outdoor cables. Aluminium and brass are also used as electrical conductors.

There are other materials in which all the electrons are so tightly bound to their parent atoms that free electron flow is impossible; such a material is known as an *insulator*. Plastics such as polyvinyl chloride (PVC), or impregnated paper are commonly used as insulators. Other insulators are

less commonly used, for example the mineral magnesium oxide, or vulcanised rubber (VRI). VRI has now been virtually superceded by PVC insulation.

10.3 Insulated conductors
Conductors are made up in cable form which is convenient for effecting a connection between the electrical supply source and the equipment making use of it. There is a great variety of cables available, but in general a cable consists of a core of small metal wires twisted together, which *conducts* the electricity, surrounded by an *insulating* material, the function of which is to confine the electricity to the conductor.

Cables may have one or more cores as described above, or may have alternate conductor and insulation layers. The number of wire strands and cable size vary depending upon the circuit requirements. Figure 10.5 shows a small selection of cable sections.

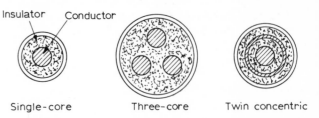

Fig. 10.5 Cable sections

Cable colours
The IEE Regulations specify colour codes for cables. In a four-wire supply system used to give a three-phase and single-phase a.c. electricity supply, for example (as illustrated in fig. 10.6), the cable colours are:

first phase   – red
second phase – yellow
third phase   – blue
neutral   – black

Cable or wiring systems
There are many different systems used for installing wiring circuits, and a specialist textbook (such as *Modern Wiring Practice*, by W. E. Steward) should be consulted for detailed information. We shall have space to consider only two systems, viz. *conduit systems* and *bus-bar systems*, the latter being dealt with in section 10.4.

Conduit systems (see BS 31 'Steel conduit and fittings for electrical wiring').

A conduit is a channel, or more often a tube, of light gauge into which cables are drawn; hence the conduit affords a cover and protection for the cable. Conduit may be metallic (steel, aluminium, or copper) or non-metallic (plastics), the tube and fittings being either screwed or, exceptionally, plain. The installation and connecting up of the conduit and fittings into a complete circuit is identical in principle to the installation of a pipework system, described in Chapter 9. The fittings are similar in shape and purpose, but many conduit fittings are made with an 'inspection cover' which is easily removed to give access to the wire inside the conduit (see fig. 10.7). A conduit system for each circuit is erected completely before any cables are drawn into the tubes, and the inspection openings mentioned above are used in this 'threading' operation.

Conduit is sometimes installed by fastening the tubing at convenient points to a wall or roof truss, etc., using 'saddle' clips, this being known as a *surface system*. Alternatively, the conduit system may be concealed in concrete, or under a wooden floor, this being known as a *concealed system*.

Where a conduit run is taken to a switch position, say, it must terminate with a metal box or into a recess. This is shown in fig. 10.7; the metal box which houses a switch may be fastened to a wall (surface system) or recessed into the wall (concealed system).

BS 1710 recommends that all conduit tubing in a building be painted *orange*, so that it cannot be confused with other pipe circuits.

Cable joints and connections

A cable may have to be *joined* to another cable or *connected* to a screw terminal, say, and in a conduit system this is done in a metal box (see fig. 10.7). The Factory Electricity Regulations state that, 'Every electrical joint and connection shall be of proper construction as regards conductivity, insulation, mechanical strength, and protection.' Figure 10.8 shows the more common cable joints and connections.

a) SOLDERED JOINTS The wire cores of the two cables are twisted together and soldered.

b) MECHANICAL CONNECTORS Small cables, up to 6mm²

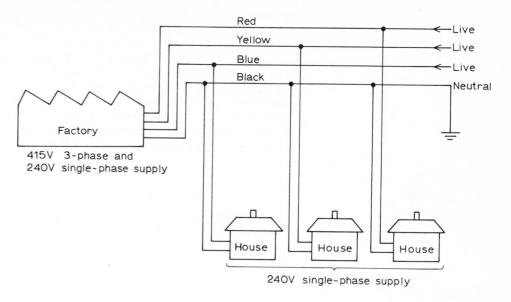

Fig. 10.6 Four-wire supply system

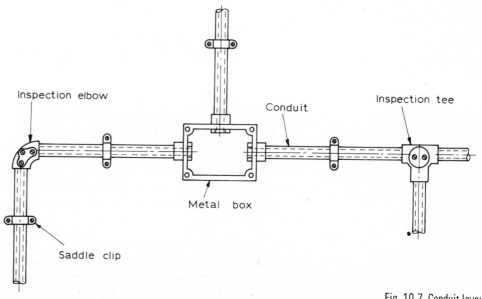

Fig. 10.7 Conduit layout

111

area, may be connected to an insulating block by means of grub screws. These blocks, usually of plastics material, are obtainable in units which are easily cut to give the required number of connections.

c) CRIMPED CONNECTORS These connectors, which may have a closed end (as drawn) or an open end, are placed over the cable conductor core and are closed by crimping to give a firm joint. They may be used to connect a cable to a terminal screw, or cables to each other.

In each of the above cases, insulator and conductor ends should be protected by wrapping around with insulation tape.

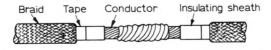

(a) Soldered joint

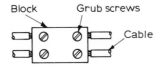

(b) Mechanical connector

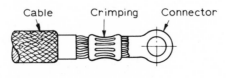

(c) Crimped connector

Fig. 10.8 Cable joints and connections

10.4 Non-insulated conductors
Non-insulated conductors may be used for such purposes as collector wires for cranes, bus-bar systems, rising mains, etc. *Bus-bars* are used extensively in industry for internal

power distribution from the main source of supply to machines on the shop floor of the factory. Vertical rising mains are similar to bus-bars, being used in domestic installations such as multi-storey flats.

Bus-bar systems
The IEE Regulations recommend that bare conductors used for rising mains or bus-bars should be installed only in places which are inaccessible to unauthorised persons. In the case of bus-bars used in factories for three-phase power distribution, this place is usually overhead.

Bus-bars are made of round- or square-section copper mounted on insulators and enclosed in lengths of steel trunking. The sections of trunking are bolted together and are fastened to the roof trusses, so that they form a continuous bus-bar system along the whole length of a machine shop. Tap-off points are provided at intervals, usually 1 m apart. Plug-in fuse-boxes (containing three fuses) are used at the tap-off points, to make connection with the three live bus-bars. Connection between the bus-bar fuse-box and the machine-motor terminal box is made by cable, usually in solid or flexible conduit. Good earth continuity is provided by an external copper earth link.

10.5 Electrical assembly drawings
We will complete this chapter by giving some examples of electrical drawings of relatively simple electrical assemblies.

Figure. 10.9 shows a circuit for a bell which may be operated by any one of three bell-pushes wired in parallel.

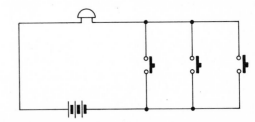

Fig. 10.9 Bell-circuit diagram

Figure 10.10 shows a circuit for the rotor-starting method used on three-phase, slip-ring induction motors. This type of motor has a wound stator and a wound rotor, with the wires brought out to three slip-rings on which bear carbon brushes. The motor is started by switching on the voltage to the stator by means of a three-pole switch and fuses.

Figure 10.11 is a block diagram of a resistance spot-welder. The welding current is conducted to the workpiece through copper electrodes which effect the weld. The transformer steps down the input to a low voltage heavy current, as selected by the tappings of the primary windings. This variable, combined with variable times as set on the timer, gives welds of different characteristics as required to suit workpiece conditions.

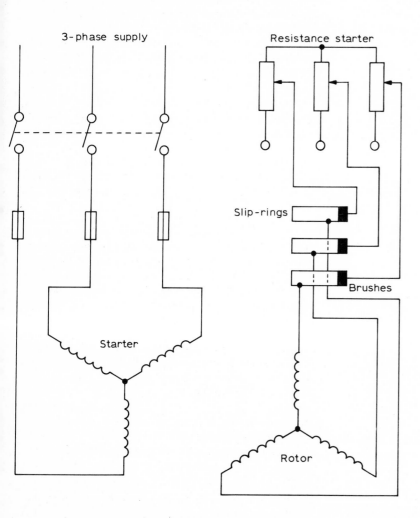

Fig. 10.10 Motor-starting-circuit diagram

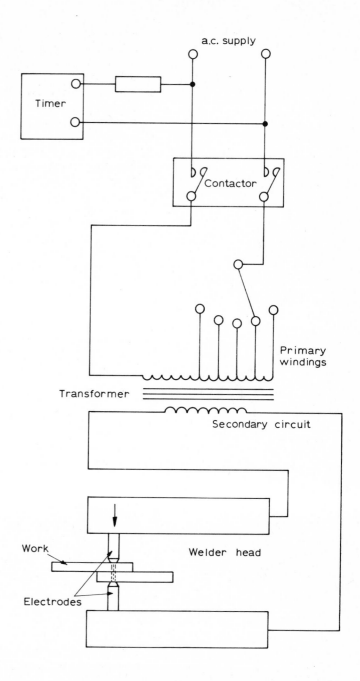

Fig. 10.11 Spot-welding-machine block diagram

113

1 a) Explain the meaning of the following electrical terms: (i) three-phase supply, (ii) bus-bar trunking system, (iii) trip switch.

b) Using conventional symbols, draw a series/parallel circuit so that two banks of four lamps may be supplied from the same main and used independently or together as required. NCTEC

2 The terminal box in fig. 10.12 is to accommodate cables from the motor windings. The terminal connections would be arranged on a board supported by the rear flanges of the box. External VRI cables would be led to the terminals through the bottom of the box.

a) Name two materials which could be used for the terminal board, and explain the mechanical and electrical properties required.

b) Explain the term 'VRI cables' and the materials used to construct the cables. Name the properties required in the core and covering, and state to what extent the properties are limited in this type of cable.

c) Make neat sketches of *either* (i) a typical terminal connection for the cables to be fastened to the board, *or* (ii) a suitable gland for the external cable entry through the bottom of the box. NCTEC

3 a) What is the meaning of the British Standard symbols or abbreviations shown in fig. 10.13?

b) Draw the British Standard symbols for the following: (i) a bearing on a shaft, (ii) a fillet weld, (iii) an a.c. bell, (iv) a non-return valve, (v) a pipe with screwed male and female joints. UEI

Fig. 10.13

(i) M/C   (ii) $\frac{63}{16}$ ALL OVER   (iii) ▭

(iv)   (v)

4 a) List three conducting materials and three insulating materials which are used in electrical industries. Select one material from each group, and give the properties which make it suited to a particular use.

b) Write notes on 'safety precautions necessary when undertaking electrical repairs to a machine'. NCTEC

5 a) Symbolic notation is adopted by BSI for welds, electrical diagrams, pipe layouts, and standard conventions. Discuss the advantages to be gained in such cases, and draw attention to the problems that can occur with the use of 'non-standard' conventions.

b) What do the symbols shown in fig. 10.14 represent? NCTEC

6 In a materials test, it is necessary to know when a specimen has fractured. For this purpose, an electrically operated signalling system is incorporated in the test-rig, and a further lamp and buzzer in the office. These are to be operated when the specimen fractures and closes a microswitch (push-button type), thus completing the circuit.

A second switch is to be included in the office, to cancel the alarm after the end of the test has been noted. The

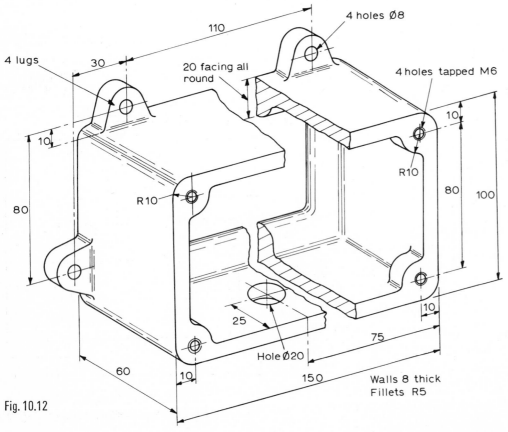

4 lugs
4 holes Ø8
110
30
20 facing all round
4 holes tapped M6
10
R10
R10
10
80
80
100
25
Hole Ø20
10
60
10
75
150
Walls 8 thick
Fillets R5

Fig. 10.12

circuit is to be connected to a battery of three cells, each of e.m.f. 2 V.

a) Using symbols for general electrical purposes, as recommended in BS 108 as amended by BS 3939, draw a suitable circuit diagram for the signalling system.

b) Make a neat sketch of a crimped terminal connector for the low-voltage cable in this system.   EMEU

7 a) Three lamps are to be installed in parallel in a lighting circuit, each with its own switch. Figure 10.15 shows (i) the lamps, (ii) the switches, (iii) the three-plate ceiling roses, and (iv) the source of supply from the distribution board in tough-rubber sheathed wiring. Copy the diagram and complete the circuit, using the relevant conventional symbols.

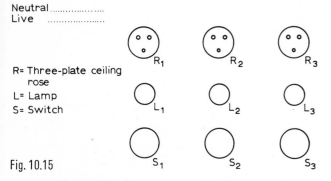

R= Three-plate ceiling rose
L= Lamp
S= Switch

Fig. 10.15

b) Figure 10.16(a) shows an all-insulated four-way terminal box. It is to be used to connect four wires in tough-rubber insulation in order to complete the circuit shown in fig. 10.16(b). Copy the terminal-box diagram, and show how the connections would be made.   NCTEC

8 Figure 10.17 shows a block diagram for an electric-bell circuit.

a) Draw this circuit as a schematic wiring diagram, using the BS symbols.

b) If these are to be mounted on a panel on a brick wall, make a sketch of a suitable arrangement.   UEI

mak

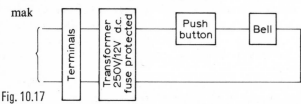

Fig. 10.17

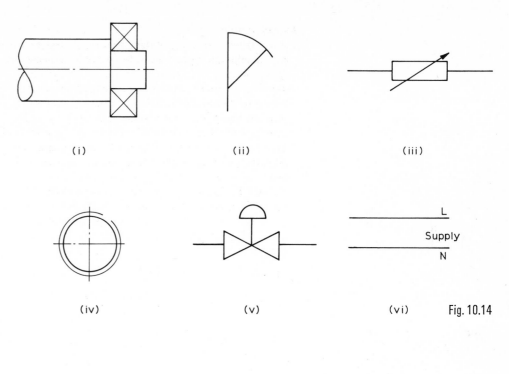

(i)   (ii)   (iii)

(iv)   (v)   (vi)   Fig. 10.14

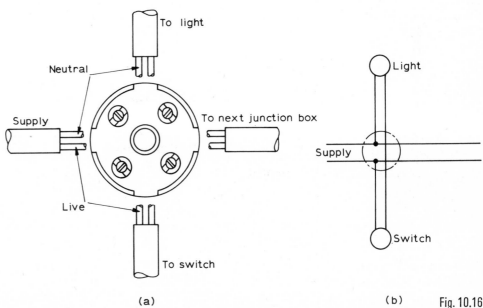

(a)   (b)   Fig. 10.16

115

9 a) Connections between electrical conductors can be made either mechanically or by soldering. Give an example of each method, and state the advantage and disadvantage.
b) The diagram shown in fig. 10.18 illustrates (i) a transformer with 8 volt and 4 volt tappings, each tapping being fitted with a cartridge fuse, (ii) a selector switch, (iii) four bell-pushes, (iv) two bells.

An 8 volt circuit is required such that three of the bell-pushes operate the day bell when the selector switch is thrown: the remaining bell-push will operate the night bell.

Copy the diagram given, and complete a schematic layout of the circuit to function as described.　　NCTEC

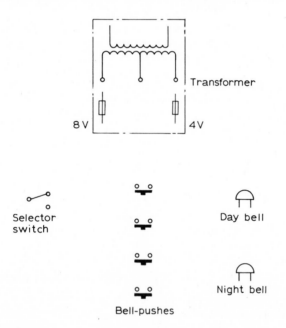

Fig. 10.18

116

Index